Maths for A Level Biology
A Course Companion

Updated Edition

Marianne Izen

Supports A Level Biology courses from AQA, Pearson, OCR, WJEC, CCEA, the International Baccalaureate and the Cambridge Pre-U

Illuminate
Publishing

Published in 2016 by Illuminate Publishing Ltd, P.O Box 1160,
Cheltenham, Gloucestershire GL50 9RW

First edition published by Illuminate Publishing in 2014

Orders: Please visit www.illuminatepublishing.com
or email sales@illuminatepublishing.com

British Library Cataloguing in Publication Data

A catalogue record for this book is available from the British Library

ISBN 978-1-908682-34-5

Printed by Cambrian Printers
01.16

The publisher's policy is to use papers that are natural, renewable and recyclable products made from wood grown in sustainable forests. The logging and manufacturing processes are expected to conform to the environmental regulations of the country of origin.

Editor: Geoff Tuttle
Cover and text design: Nigel Harriss
Text and layout: GreenGate Publishing Services, Tonbridge, Kent

Photo credits

Cover © Science Photo Library; **6** © Matej Kastelic/Shutterstock; **p24** © Istock.com/Henrik_L; **p26** © Istock.com/BeholdingEye; **p31** © Istock.com/Angelika Stern; **p32** © Eric Isselee/Shutterstock; **p44** © Istock.com/Petr Malyshev; **p45** © Istock.com/Lisay; **p60** © Istock.com/Iurii; **p64** © Richard Southall/Shutterstock; **p72** © Eric Isselee/Shutterstock; **p72** © Istock.com/GlobalP; **p72** © Maria Pawlak/Shutterstock; **p73** © skydie/Shutterstock; **p74** © Potapov Alexander/Shutterstock; **p74** © doomu/Shutterstock; **p77** © Patryk Kosmider/Shutterstock; **p77** © Cosmin Manci/Shutterstock; **p77** © Dimarion/Shutterstock; **p83** © InavanHateren/Shutterstock; **p84** © Tatiana Volgutova/Shutterstock; **p87** © Istock.com/Czhan; **p91** © NADKI/Shutterstock; **p91** © ravl/Shutterstock; **p93** © Shcherbakov Ilya/Shutterstock; **p94** © Photok.dk/Shutterstock; **p94** © Istock.com/bergamont; **p95** © schachspieler – Fotolia.com; **p95** © morrison/Shutterstock; **p96** © anyaivanova/Shutterstock; **p96** © Africa Studio/Shutterstock; **p96** © Tsekhmister/Shutterstock; **p97** © Ratikova/Shutterstock; **p97** © Paul Staniszewski/Shutterstock; **p100** © Istock.com/JJMaree; **p114** © Unholy Vault Designs/Shutterstock; **p122** © Istock.com/Linda Steward; **p126** © Garry DeLong – Fotolia.com; **p133** © Istock.com/Alasdair Thomson; **p133** © Istock.com/Suzifoo; **p135** © Eugenio Chelli – Fotolia.com; **p135** © Istock.com/philipatherton

Acknowledgements

The author and publisher wish to thank the following for their valuable contributions to this book:

Dr Colin Blake
Elizabeth Humble
Rachel Knightley
Dr Meic Morgan
Cara Patel

Contents

| Introduction | 6 |
| How to use this book | 7 |

In the contents below, the bullet points show where biological concepts are used to illustrate the mathematics.

① Numbers   9

1.1 Arithmetic   9

- **1.1.1 Addition**
  - Diffusion distance
  - Water potential
  - Tissue fluid movement
  - Solute potential
- **1.1.2 Subtraction**
  - Energy yield in farming
  - GPP and NPP
- **1.1.3 Multiplication**
  - Energy content of food
- **1.1.4 Division**
  - Body mass index
  - Chromatography and R_f
- **1.1.5 Estimation**

1.2 Using the calculator   15
  - Bacterial culture with antibiotics

1.3 The order of operations   16

1.4 Powers, indices and standard notation   16

- **1.4.1 Indices**
  - Indices and the genetic code
  - Indices, gametes and zygotes

- **1.4.2 Powers of 10**
  - Expressing reaction rate
- **1.4.3 Negative indices**
- **1.4.4 Negative indices in units**
- **1.4.5 Indices and arithmetic**
  - Serial dilution
- **1.4.6 Using logarithms**

1.5 Ratios   21
  - Body proportions and age
  - BOD in different water types
  - Respiratory quotient

1.6 Fractions   24

1.7 Decimals   24

- **1.7.1 Decimal notation**
- **1.7.2 The number of decimal places**
  - Mark and recapture calculation
- **1.7.3 Rounding up and down**
- **1.7.4 Significant figures**

1.8 Negative numbers   27
  - Embryonic development
  - Water relations in plant cells

Test yourself 1   30

② Processed numbers   32

2.1 Percentage   32

- **2.1.1 Per cent calculation**
  - Bird plumage
- **2.1.2 Per cent increase and decrease**
  - Mass change in rodents
- **2.1.3 Per cent frequency**
  - Point frame quadrat
  - Gridded quadrat
- **2.1.4 Per cent area cover**
  - Gridded quadrat
- **2.1.5 Percentage error**

2.2 Proportion   35

- **2.2.1 Surface area and volume**
  - Diffusion
  - Organisms as cubes
  - Organisms as spheres
  - Organisms as cylinders
- **2.2.2 DNA replication**

2.3 The concentration of a solution   42

- **2.3.1 Per cent concentration**
- **2.3.2 Using moles**
- **2.3.3 Molarity**
- **2.3.4 Hydrogen peroxide**

2.4   Biotic indices   44
    2.4.1   Lincoln index
    2.4.2   Simpson's index
    2.4.3   Heterozygosity index
2.5   The haemocytometer   46

2.6   Joints as levers   47
2.7   Unfamiliar mathematical expressions   49
    ▪ Assimilation number
    ▪ Energy efficiency
Test yourself 2   51

❸ Graphs   52

3.1   Axes   52
3.2   Axis scales   52
3.3   Types of data   54
    3.3.1   Categorical
      ▪ Glucose content of fruit
    3.3.2   Discrete
      ▪ Petal number
    3.3.3   Continuous
      ▪ Human height
3.4   Population pyramids   56
3.5   Line graphs   57
    3.5.1   Axis choice
    3.5.2   Axis labels
    3.5.3   Scale choice
    3.5.4   Points
    3.5.5   Line
    3.5.6   Key

3.6   Interpreting graphs   60
    3.6.1   Describing graphs
      ▪ Rate of reaction
      ▪ Bloodworm distribution
      ▪ Population curves
    3.6.2   Numerical analysis of graphs
      ▪ Substrate concentration and rate of reaction
      ▪ Compensation point
      ▪ Oxygen dissociation curves
      ▪ Rate of sugar formation
      ▪ Percentage increase sugar formation
      ▪ Percentage decrease internode length
    3.6.3   Rates of reaction
      ▪ Mass of sugar produced
      ▪ Mass of protein digested
    3.6.4   Population growth curves
    3.6.5   Spirometer traces
    3.6.6   ECG traces
    3.6.7   Kite diagrams
3.7   Pie charts   75
    ▪ Percentage cover herbaceous plants
    ▪ Cell cycle
3.8   Nomograms   75
    ▪ Protein digestion
Test yourself 3   76

❹ Scale   79

4.1   Units   80
4.2   Converting between units   80
4.3   How to indicate scale   81
4.4   Microscope calibration   81

4.5   Magnification   84
4.6   Area   85
4.7   Volume   86
4.8   Constructing ecological pyramids   87
Test yourself 4   89

❺ Ratios and their use in genetics   90

5.1   Blending inheritance   90
5.2   Monohybrid inheritance   91
    5.2.1   Crosses and offspring
    5.2.2   Monohybrid crosses
    5.2.3   Partial dominance
5.3   Dihybrid crosses   96
    ▪ Ratios and numbers of offspring

5.4   Theoretical ratios and real life   100
5.5   Non-Mendelian ratios   102
    5.5.1   Lethal recessives
    5.5.2   Epistasis
    5.5.3   Linkage
    5.5.4   Sex linkage
Test yourself 5   108

6 **The Hardy–Weinberg Equilibrium**   **110**

    Test yourself 6   112

7 **Statistics**   **113**

7.1	Data	113	7.8 Level of significance	123
7.2	Sampling	113	7.9 Confidence limits	123
7.3	Probability	114	7.10 Degrees of freedom	123
	▪ Seed germination		7.11 Fitting confidence limits to the mean	124
7.4	Averages	114	▪ Lengths of earthworms	
7.4.1	**Arithmetic mean**		7.12 Ranking	126
7.4.2	**Median**		▪ Leaf miner tunnel length	
	▪ Leaf length		▪ Human body mass	
7.4.3	**Mode**		▪ Acorn length	
	▪ Spots on ladybirds		7.13 One-tailed and two-tailed tests	127
7.5	Distributions	115	7.14 Correlation	128
7.5.1	**Normal**		▪ Dogs, their owner and their tails	
	▪ Human height		7.15 Choosing a statistical test	129
7.5.2	**Negative skew**		7.16 The Spearman rank correlation test	131
	▪ Length of fish		▪ Light intensity and leaf internode	
7.5.3	**Positive skew**		length	
	▪ Swimming speed		7.17 The Pearson linear coefficient test	134
7.5.4	**Bimodal**		▪ Pollen tube length and time since	
	▪ Length of fish		germination	
7.6	Variability	117	7.18 The Mann-Whitney U test	137
7.6.1	**Range**		▪ Water shrimp number and	
7.6.2	**Standard deviation**		substrate type	
	▪ Width of middle finger		7.19 The t test	139
7.6.3	**Variance**		▪ Mayfly nymph number and oxygen	
7.6.4	**Standard error**		concentration	
7.6.5	**Percentiles and quartiles**		7.20 The χ^2 test	141
7.7	Making a null hypothesis	122	▪ Seed colour and Mendelian	
	▪ Phosphate ions and stone fly nymphs		inhcritance	
	▪ Mendelian genetics		Test yourself 7	145

Quickfire answers   **148**

Test yourself answers   **152**

Glossary   **159**

Specification map   **163**

Index   **167**

Introduction

This is not a Mathematics text book. Nor is it a Biology text book. It is a book about the use of some mathematical concepts in Biology. This is a book that will explain to you how some numerical, geometrical and statistical ideas are used in post-16 Biology courses, and how to approach calculations when you are taking these courses and sitting their examinations.

If you are one of the many students at this level who are science students, you may also be taking Chemistry, Physics, Maths, Geography or Psychology. This book is for you. Some of its content may be familiar but explanations in the context of biology will ensure that you use the mathematics as a tool for understanding more about living things.

On the other hand, perhaps you are one of the many Biology students principally studying arts or humanities. You may have breathed a sigh of relief at the end of your last GCSE Maths exam, only to receive a nasty shock in Year 12 Biology lessons. This book is most definitely for you. It will explain the concepts that you need using directly relevant examples and will explain to you how to avoid common misunderstandings when making calculations. It will show you how to decode examination questions to find exactly what biological question is being asked and will demonstrate how to put the biological problem into mathematical form to find an answer.

You will be helped throughout this book with Pointers, Quickfire tests, worked examples and Test yourself questions similar to those you may find in an examination. The answers and the glossary at the back of the book will let you check your progress.

This is Biology, not Mathematics. In Biology A Level examinations, whichever board you are sitting, a minimum of 10% of the marks are for mathematics. So it is important not to be one of those students who sees numbers, takes fright and moves on to the next question. Read the question as many times as you need to understand what it is asking you. The meaning will eventually become clear. Then apply the logic you have gleaned from this book and you will never fear a calculation again.

How to use this book

This book is a combination of mathematics and biology. There is no logical way to present the information in order of difficulty because everything is inter-related. However, each chapter deals with a major aspect of mathematics as it is used in biology. Basic concepts such as arithmetic and the relevant geometry are towards the beginning and the more involved processes such as the use of quadratic equations and of statistical tests come later.

There are several ways this book will help you

1. Using the Contents, Index and Specification map, you will be able to locate what you need. The Contents shows the mathematics covered in each chapter and shows how it is used in various biological contexts. It can be used in conjunction with the Specification map on pages 163–166. This Specification map shows the mathematics requirements stated in each examination board's specifications.

2. Various terms throughout this book have been highlighted. These all appear in the **Glossary** on pages 159–162, which gives you their definitions.

3. Each chapter has **Pointers** which state important ideas and it may be useful for you to memorise these.

$\gg$ *Pointer*

4. There are **Quickfire questions** for you to check that you follow the concepts explained, although some may be quicker than others to answer. The answers are all given so that you can check your understanding. If you get them wrong, go back over the text and try again.

quickfire $\gg$

5. **Worked examples** are given throughout the text using biological scenarios, in the way that they may be presented to you in examination questions. Work through these, paying attention to the way they are set out. Here's an example:

This is how it works:

Here is a calculation: $4 \times (2^2 + 3) \div 4 + 7 - 11$

First deal with the powers:	$= 4 \times (4 + 3) \div 4 + 7 - 11$
Then remove brackets	$= 4 \times 7 \div 4 + 7 - 11$
Then multiply:	$= 28 \div 4 + 7 - 11$
Then divide:	$= 7 + 7 - 11$
Then add:	$= 14 - 11$
Then subtract:	$= 3$

---

6   Each chapter ends with **Test yourself questions**. These are in the style of examination questions, and worked answers are given for all of them. Be sure that when you write your own answers, they provide every step of the logic and that they are written in a clear sequence of accurate mathematical statements. That way, your examiner can see that you know what you are doing and that you understand the mathematics behind the biology.

Test yourself

1 If 34% of the bases in a molecule of DNA are thymine, what percentage of the bases are guanine?

2 At maximum inspiration, the air pressure in the alveoli is 0.30 kPa below atmospheric pressure.

At maximum expiration, the alveolar pressure is 0.29 kPa above atmospheric pressure. Calculate the difference in alveolar pressure during one cycle of breathing.

3 One complete cardiac cycle lasted from 0.50 seconds to 1.34 seconds after measuring began. Use this information to calculate the heart rate.

Chapter 1

Numbers

1.1 Arithmetic

Calculations in Biology are a means to an end and so you may be asked to add, subtract, multiply or divide to find out something about a biological situation. Here is a summary of the arithmetic you knew for GCSE, as applied to Biology. As you can see, the sums are easy but you have to understand the biology to know what to do.

1.1.1 Addition

Adding numbers is straightforward. The biological skill is in deciding what numbers have to be added.

The mathematical skill is in knowing that adding two positive numbers is simple addition [$a + b = x$], but that adding a positive to a negative number is equivalent to a subtraction [$a + (-b) = a - b = y$] and that adding two negative numbers means you subtract them both [$(-a) + (-b) = -a - b = z$]. Here are some examples.

Adding positive values: what is the minimum distance a molecule of carbon dioxide diffuses to move from the plasma to the air in the alveolus?

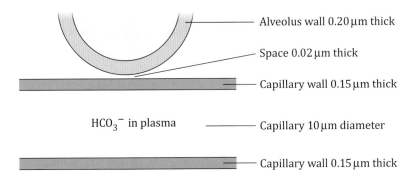

The diagram shows two capillary walls and the wall of the alveolus, with a small gap between. The minimum distance the molecule moves will be across the capillary wall (0.15 μm), across the gap (0.02 μm) and across the alveolus wall (0.20 μm).

> Add the values together to give the distance: 0.15 + 0.02 + 0.20 = 0.37 μm

Questions in examinations generally say, 'Show your working', so you must write out the whole sum, as shown here. If you do not, even if you get the right answer, you may not be awarded full marks. But if you get the wrong answer, as long as your working is logical, you may be credited for that. So showing your working is an insurance policy.

quickfire >> 1.1

The egg and sperm of a fox each have 17 chromosomes. What is the diploid number of a fox?

>> **Pointer**
Show your working.

1.2

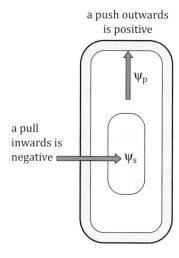

≫ Pointer

Never forget the units.

a push outwards is positive

ψ_p

a pull inwards is negative

ψ_s

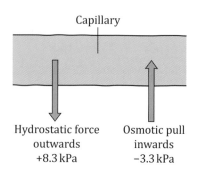

Capillary

Hydrostatic force outwards +8.3 kPa

Osmotic pull inwards −3.3 kPa

If you do not include the units, even if the calculation is correct, you will not get full marks. This is because this question is about Biology not Mathematics and so such details are significant. Sometimes, the units may be written at the end of a dotted line where your answer should go. If you write the units in addition, you will not be penalised. Twice is better than not at all.

Adding positive and negative values

a) Plant cell water relations

The diagram on the left shows a plant cell with its pressure potential (ψ_p) and solute potential (ψ_s) indicated.

The pressure potential can be thought of as a push outwards and so it has a positive value. The solute potential can be thought of as an osmotic pull inwards and so it has a negative value.

An examination question may be phrased like this: Calculate the water potential of the cell if its pressure potential is 300 kPa and its solute potential is −750 kPa. Show your working.

This is not a maths test, so you know that the actual calculation is easy. What is being tested is your understanding of water relations of a cell. You are expected to know that the water potential of a cell is the balance of the pressure and solute potentials and the sum of the two will tell you which way water will move. This means you must add a positive to a negative value, which is the same as subtracting the negative value from the positive one, as shown in the box below.

Always start with the equation:

water potential	=	pressure potential	+	solute potential
ψ_c	=	ψ_p	+	ψ_s

Then substitute in the numbers: ψ_c = 300 + (−750)

Then give the calculation: ψ_c = 300 − 750 = −450 kPa

Water relations of plant cells are discussed further on pages 28–29.

b) Plasma and tissue fluid

You may also find the need to add positive and negative numbers when considering the exchange between plasma and tissue fluid in a capillary bed. The hydrostatic pressure of the blood pushes outwards and has a positive value. The osmotic potential of the plasma pulls inwards and has a negative value. The direction of movement of fluid depends on the balance of the two:

$$\text{Resultant force} = 8.3 + (-3.3)$$
$$= 8.3 - 3.3$$
$$= 5.0 \text{ kPa}$$

This resultant is positive, which means it is a push outwards, and so liquid leaves the capillary to form tissue fluid which bathes the cells.

Adding two negative numbers

You may be asked to calculate the solute potential of a solution with more than one solute. The total solute potential is the sum of the contributions from each solute, so you add them.

Imagine a solution containing 0.1 mol dm^{-3} glucose and 0.2 mol dm^{-3} sucrose. The solute potential due to the glucose is −2.6 kPa and the solute potential due to the sucrose is −5.3 kPa.

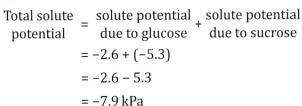

$$= -2.6 + (-5.3)$$
$$= -2.6 - 5.3$$
$$= -7.9 \text{ kPa}$$

 Pointer

Calculations look better if each equals sign is directly below the one in the line above.

1.1.2 Subtraction

Decoding a question tells you what you must do with the data, and if it is not immediately clear, keep on reading it over and over until it makes sense. Eventually it will. Here is an example where you are asked to find the net energy yield per hectare with different farming methods.

This question gives a sample calculation, so when you think you know what you have to do, make sure your method gives the same answer as the one given.

You should study all the data carefully so you know exactly what the question is telling you. This tells you how much energy you have to put in as a hunter-gatherer or as a wheat or dairy farmer, and how much energy you get in the food produced.

 Pointer

Check whether your result is positive or negative and that you have not forgotten a negative sign if it is needed.

Find the net energy yield per hectare from wheat monoculture and from dairy farming.

Type of farming	Energy per hectare / arbitrary units		
	Input	Actual yield	Net yield
Hunting and gathering	0.40	2.80	2.40
Wheat monoculture	15 500	50 000	
Dairy	27 000	10 000	

'Net yield' means the difference between the two so: net yield = actual yield − input

For hunting and gathering,

net yield = 2.80 − 0.40 = 2.40 arbitrary units, as given in the table.

Using the same method, for wheat monoculture.

net yield = 50 000 − 15 500 = 34 500 arbitrary units

For dairy farming,   net yield = 10 000 − 27 000 = − 17 000 arbitrary units

Don't omit the negative sign. This tells us that with dairy farming, you put in more energy than you get out. You may be asked to link the result of this calculation with the cost of producing food using current agricultural techniques as the world's population increases.

A commonly asked subtraction sum relates to primary productivity. The gross primary productivity (GPP) of an ecosystem represents the energy fixed by autotrophs in a given area in a given time. It is sometimes defined as the rate at which energy is converted by photosynthesis and chemosynthesis into organic substances. It is generally quoted as kJ/hectare/year or kJ/m^2/ year. This is the same as writing kJ ha^{-1} y^{-1} or kJ m^{-2} y^{-1}. The net primary productivity (NPP) is the energy remaining after the producer has respired carbohydrates made in photosynthesis. This is the energy that is available to the next trophic level.

» Pointer
NPP = GPP – R

» Pointer
Check that the answer to your calculation makes sense.

» Pointer
Line up the numbers beneath the relevant words of the equation so that your examiner will know that you understand what you are doing.

The sum you are asked to do uses the formula NPP = GPP – R where R represents the energy no longer available when the producer is eaten, due to its respiration.

Actual values of two of these may be given to allow the third to be calculated or you may be expected to find the values from an energy flow diagram.

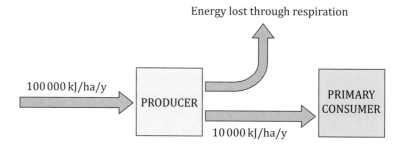

From the diagram, energy lost through respiration
R = GPP – NPP

= 100 000 – 10 000

= 90 000 kJ/ha/y

When you make a calculation, consider if your answer makes sense in the light of your theoretical knowledge. You may remember that about 90% of the energy entering a trophic level is lost through respiration and this answer is consistent with that information.

In a Sankey diagram the width of the arrows is proportional to the energy flow represented:

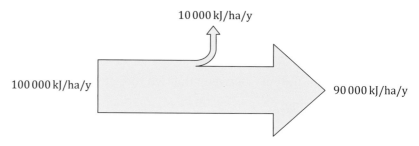

1.1.3 Multiplication

Simple multiplication can be used to make sure you can picture a physiological system such as breathing. In this example, you are asked to find the volume of air, in dm³, breathed per minute by a physically fit adult who breathes 14 times a minute and exchanges 400 cm³ of air with each breath.

You can use the wording of the question to produce an equation:

Volume of air breathed per minute	=	number of breaths per minute	×	volume of air exchanged with each breath
	=	14	×	400
	= 5600 cm³ = 5.6 dm³			

Multiplication is used in calorimetry, when the energy content of a food is calculated. The energy released from burning a known mass of food is absorbed by a known mass of water. The temperature rise of the water is measured. Assuming that no energy released by the food is lost from the system, and that all the energy goes into heating the water rather than its container,

$$\text{energy absorbed by the water} = \text{mass of water} \times \text{specific heat capacity of water} \times \text{temperature rise of water}$$

Considering the units,

$$\text{energy absorbed} = \text{mass (g)} \times \text{SHC}\left(\frac{J}{g\,C°}\right) \times \text{temperature rise (C°)}.$$

The units g and C° occur in the numerator and the denominator and so cancel out, leaving the units J.

> Calculate the energy content of crisps if burning 0.5 g increases the temperature of 25 cm³ water from 20 °C to 50 °C.
>
> Energy released from 0.5 g
>
> = (mass water × specific heat capacity of water × temperature rise) J
>
> = 25 × 4.2 × (50 − 20) J
>
> = 3150 J
>
> Energy from 1 g $= \dfrac{3150}{0.5} = 6300\,J$
>
> ∴ Energy released = 6.3 kJ/g

1.1.4 Division

Some calculations ask you to divide. You can represent this as the division sign, e.g. 6 ÷ 3 = 2, or as a fraction e.g. $\frac{6}{3} = 2$.

1 You may, for example, be asked to calculate someone's body mass index, given their mass and height. You would not be expected to remember the equation, so all you have to do is substitute in correctly.

E.g. Calculate the BMI of an adult who is 1.70 m tall and who weighs 65.03 kg using the equation

$$BMI = \frac{\text{mass / kg}}{(\text{height / m})^2}$$

> Start by rewriting the equation: $\quad BMI = \dfrac{\text{mass / kg}}{(\text{height / m})^2}$
>
> Substitute in the figures: $\quad BMI = \dfrac{65.03}{1.70^2}$
>
> Give each stage of the calculation: $\quad BMI = \dfrac{65.03}{1.70 \times 1.70}$
>
> $\qquad = \dfrac{65.03}{2.89}$
>
> $\qquad = 22.50$

quickfire 1.6

A plant cell in tissue culture has a radius of 34 µm. Calculate its volume in µm³ using the expression for the volume of a sphere:

volume of a sphere = $\frac{4}{3}\pi r^3$, where r = cell radius.

Divide by 10⁹ to give the answer in mm³.

State the number of decimal places to which your answer is given.

> **Pointer**
> Make sure you have given your answer in suitable units.

quickfire 1.7

The respiratory quotient, RQ, for an organism is given by the equation:

$RQ = \dfrac{\text{volume of } CO_2 \text{ emitted}}{\text{volume } O_2 \text{ absorbed}}$

Calculate the value of RQ for an earthworm that takes in 20 mm³ of oxygen for every 16 mm³ carbon dioxide it releases.

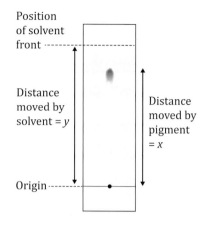

Position of solvent front ┄┄┄

Distance moved by solvent = y

Distance moved by pigment = x

Origin ┄┄┄

As this is Biology and not Maths, you may be asked to evaluate this person's BMI, given that a BMI of below 20 is considered underweight and that a BMI of over 25 is described as overweight. You could deduce that this person's BMI is within the normal range.

2   Calculating R_f for the photosynthetic pigments is another situation where you will encounter division. A standard lab task is to separate pigments by chromatography. You may produce a chromatogram as shown on the left.

For each pigment, $R_f = \dfrac{x}{y}$.

The actual value for a particular pigment depends on the solvents used for its extraction and for running the chromatogram.

1.1.5 Estimation

When people use a calculator, they often press the wrong button or miss a decimal point, but because the calculation has been done on a calculator, they assume it is the right answer. It is always useful to estimate an answer before you do your working. That way, you will know if your calculated answer is likely to be correct. Estimating requires you to be able to add and subtract and to know your tables. This is why some people think that estimating is a primary school skill. You will find it very useful at this level too.

Before you start, round numbers up or down to the nearest 1 or 10 or whatever is appropriate. Use standard notation when numbers are larger than 100. Here are some examples:

Adding

a)   6.05 + 8.98

As these numbers are under 10, round to the nearest 1, and the sum becomes 6 + 9.

6 + 9 = 15. Your calculator will give the answer 15.03.

b)   If the numbers are very large, you can combine rounding with using standard form, e.g.

$345678 + 123456 \approx (3 \times 10^5) + (1 \times 10^5) = 4 \times 10^5$

Your calculator would give the answer 469134.

Subtracting

a)   29 – 11

As these are 2-digit numbers, round to the nearest 10 and the sum becomes 30 – 10

30 – 10 = 20. Your calculator will give the answer 18.

b)   Here is an example with a very large number, using rounding and indices. Remember to have the same powers of 10 for each number.

$876543210 - 7654321 \approx (9 \times 10^8) - (0.07 \times 10^8) = 8.03 \times 10^8$

Your calculator would give $868888889 = 8.69 \times 10^8$

Pointer

If your calculation is not correct but you correctly evaluate your own answer, you will only lose marks on the calculation, which was after all wrong, but not on the evaluation, as you correctly interpreted what you found.

quickpire⟫ 1.8

Estimate then calculate the answers:

a)   2 + 11
b)   3.2 + 8.9
c)   21 + 93
d)   179 + 785
e)   2698 + 7859

quickpire⟫ 1.9

Estimate then calculate the answers:

a)   19 – 2
b)   8.2 – 0.9
c)   93 – 39
d)   880 – 213
e)   6789 – 1234

Multiplying

a) 113 × 298

These are 3-digit numbers so round to the nearest 100 and the sum becomes 100 × 300

100 × 300 = 30 000. Your calculator will give the answer 33 674.

b) With very large numbers, round up or down, change to standard form and just add indices: e.g. 12345678 × 12345 ≈ $(1 \times 10^7) \times (1 \times 10^4) = 1 \times 10^{11}$

Your calculator will give the answer 1.52×10^{11}

Dividing

a) 6984 ÷ 18

Rounding the numbers gives 7000 ÷ 20 = 350. Your calculator will give the answer 388.

b) But with large numbers, you can change to standard form beginning with 1. Then, because it is division subtract the indices:

34567890 ÷ 8765 ≈ $(1 \times 10^7) \div (1 \times 10^4) = 10^3$

Your calculator will give the answer 3.94×10^3.

quickfire 1.10

Estimate then calculate the answers:

a) 9 × 8
b) 5.8 × 5.1
c) 23 × 67
d) 320 × 190
e) 31975 × 219654

quickfire 1.11

Estimate then calculate the answers:

a) 39 ÷ 4
b) 8.9 ÷ 2.9
c) 119 ÷ 6.1
d) 920 ÷ 31
e) 1234567 ÷ 11712

1.2 Using the calculator

People use their calculators far more than they need to. But if you really do need to, you must know how to use it. The commonest error that is likely to give a confusing result in Biology is not knowing when to use the equals key. It happens most frequently in calculating a mean. To find a mean, you have to add together several numbers and then divide by the number of numbers.

Let us imagine an experiment where you want to find the mean diameter of a clear zone around an antibiotic disc placed on a lawn of bacteria. The table here shows the readings.

To calculate a mean, you must add the diameters and divide by 10.

The total comes to 11+12+13+16+20+9+11+11+10+9 = 122 mm.

Because it is easy to divide by 10, you could easily say 12.2 mm. But this is a discussion about the use of a calculator. To get the correct answer, you must press = before pressing ÷, because the = adds together all the individual numbers before you divide, which is how to calculate a mean. This can be written
(11+12+13+16+20+9+11+11+10+9) ÷ 10 = 12.2 mm

If you just press ÷, without the = first, only the last number you press will be divided. That would be written 11+12+13+16+20+9+11+11+10+ (9 ÷ 10) = 113.9 mm.

Obviously 113.9 mm could not possibly be the mean of the diameters given above, so, having thought about the data before doing the calculation, you would suspect something was wrong. But students do not always bother to estimate an answer first and so they do not know what to expect. If you were that student, you would not know you had a wrong answer.

Sample number	Diameter of clear zone / mm
1	11
2	12
3	13
4	16
5	20
6	9
7	11
8	11
9	10
10	9

» Pointer

Use the = key before the ÷ key when you calculate a mean.

1.12

A man counted his pulse rate on waking every day for a week. Here are his pulse rates:

Day	Pulse rate / bpm
Sunday	64
Monday	68
Tuesday	70
Wednesday	70
Thursday	69
Friday	72
Saturday	62

a) What is his mean pulse rate on a working day?

b) Is he more relaxed at the weekend or during the week?

» Pointer

Always estimate your answer before using a calculator, so you know what to expect.

 1.13

Evaluate $3 + 9 - (3^3 - 15) \div 2 \times 3$

» Pointer

Please Bless My Dear Aunt Sally.

quickfire **1.14**

If one base codes for one amino acid, $4^1 = 4$ amino acids could be coded for.

If two bases code for one amino acid, $4^2 = 16$ amino acids could be coded for.

Work out how many amino acids can be coded for, given that three bases code for each amino acid.

1.3 The order of operations

This discussion about calculating a mean diameter suggests that when you have a calculation with several operations to perform, the order in which you do them is crucial.

After all $2^2 + 12 \div 4$ could be $(2^2 + 12) \div 4 = 16 \div 4 = 4$ or it could be $2^2 + (12 \div 4) = 4 + 3 = 7$.

There is a recognised way of doing this and the way you may have learned in primary school still works at A Level. Please Bless My Dear Aunt Sally. The order of operations is:

1 Powers   **2** Brackets   **3** Multiply   **4** Divide   **5** Add   **6** Subtract

This is how it works:

Here is a calculation: $4 \times (2^2 + 3) \div 4 + 7 - 11$

First deal with the powers:	$= 4 \times (4 + 3) \div 4 + 7 - 11$
Then remove brackets	$= 4 \times 7 \div 4 + 7 - 11$
Then multiply:	$= 28 \div 4 + 7 - 11$
Then divide:	$= 7 + 7 - 11$
Then add:	$= 14 - 11$
Then subtract:	$= 3$

You may have learned a similar approach, written as BODMAS, which means brackets, orders (e.g. powers, square roots), divide and multiply, add and subtract. The answer will be the same.

1.4 Powers, indices and standard notation

Powers and standard notation are useful tools. They are explained here, with examples that you are likely to come across when studying Biology.

1.4.1 Indices

'x' can mean any number. In Maths, you may have met the idea that $x^0 = 1$ but you are not likely to need that in Biology, though higher powers are used.

x^1 is just x, i.e. $x^1 = x$. You may meet that idea when thinking about a dilution series.

'x^2' is read as 'x squared' and it means any number multiplied by itself, in other words, x multiplied twice. So if $x = 2$, $x^2 = 2 \times 2 = 4$. The 2 in x^2 is the power, so you could call x^2 'x to the power of 2'. A mathematician might call the 2 the exponent or the index. The plural of index is indices.

In the same way, 'x^3' means any number multiplied by itself and then by itself again, so that in the calculation, the number is written three times.

So if $x = 2$, $x^3 = 2 \times 2 \times 2 = 8$. You could call x^3 'x to the power of 3' or 'x cubed'. The number 3 is the exponent or the index.

A number is in standard notation when a number from 1 to 9 is followed by 10 to a power. As an example, 5503 is not in standard notation, but 5.503×10^3 is.

Indices and the genetic code

Indices are used when considering the genetic code. When the concept arose that the sequence of bases in DNA was correlated with the sequence of amino acids in a protein, people asked the question, 'How many bases code for each amino acid?' Even before experiments were performed that verified that there is a triplet code, logical analysis had suggested the same thing. It was known that there were four different bases in the DNA code, guanine, adenine, cytosine and thymine. If one base coded for one amino acid, with four bases only four amino acids could be coded for. But 20 amino acids have to be coded for to make the proteins of living organisms, so one base coding for one amino acid is not enough. If two bases coded for one amino acid, how many could then be coded for? This is asking, how many possible combinations of two bases there are when there are four bases in all. It can be worked out like this:

» Pointer

Take care to get the number and index the right way round when you do a calculation:

$2^3 = 2 \times 2 \times 2 = 8$

but $3^2 = 3 \times 3 = 9$

quickfire » 1.15

Two genes each with two alleles can produce $2^2 = 4$ different allele combinations: AB, Ab, aB and ab.

How many combinations could be produced by three genes, A/a, B/b and C/c?

		SECOND BASE			
		G	T	A	C
	G	GG	GT	GA	GC
FIRST BASE	T	TG	TT	TA	TC
	A	AG	AT	AA	AC
	C	CG	CT	CA	CC

There are 16 possible combinations of two bases, as shown above. Another way of thinking about this is that there are $4^2 = 16$ possible combinations. But this also does not allow enough amino acids to be coded for.

The next simplest situation is to have three bases coding for each amino acid. To work out how many possible combinations of three bases there are, the table can be redrawn. It needs a third dimension but as that is not possible on the page, the third dimension is represented by each pair of bases being grouped with each of the four bases as shown on the right.

The table shows that there are $4^3 = 64$ combinations of three bases. This is the smallest number of base combinations to code for all the amino acids that appear in the genetic code, and as there are combinations in excess of what is needed, some have additional functions, such as punctuating the code.

That the simplest explantion is likely to be the correct one is one way of expressing a well-known philosophical concept of 'Occam's razor', said to have been devised by William of Ockham in the 14th century. He did not have molecular genetics in mind but his principle has many applications and provides a theoretical rationale for a triplet code.

So in summary, using indices:

one base – one amino acid codes for 4^1 = 4 amino acids

two bases – one amino acid codes for 4^2 = 16 amino acids

three bases – one amino acid codes for 4^3 = 64 amino acids

		SECOND BASE			
		G	T	A	C
		GGG	GTG	GAG	GCG
	G	GGT	GTT	GAT	GCT
		GGA	GTA	GAA	GCA
		GGC	GTC	GAC	GCC
		TGG	TTG	TAG	TCG
	T	TGT	TTT	TAT	TCT
		TGA	TTA	TAA	TCA
		TGC	TTC	TAC	TCC
FIRST BASE		AGG	ATG	AAG	ACG
	A	AGT	ATT	AAT	ACT
		AGA	ATA	AAA	ACA
		AGC	ATC	AAC	ACC
		CGG	CTG	CAG	CCG
	C	CGT	CTT	CAT	CCT
		CGA	CTA	CAA	CCA
		CGC	CTC	CAC	CCC

Indices, gametes and zygotes

When you learn about meiosis, you learn about independent assortment. This means that either of a pair of homologous chromosomes can go to either pole when the cell divides to make gametes.

- If there are 2 pairs of chromosomes (n = 2) i.e. 2 from the mother and 2 from the father, there are 2^2 = 4 possible combinations of maternal and paternal chromosomes in the gametes.
- With 3 pairs of chromosomes, there are 2^3 = 8.
- In humans there are 23 pairs of chromosomes so, there are 2^{23} = 8,388,608 possible combinations.

In summary, with n pairs of chromosomes, there are 2^n possible combinations of maternal and paternal chromosomes in the gametes.

The chromosomes of two gametes combine to make a zygote.

- When n = 2, there are 2^2 = 4 possible male gametes and 4 possible female gametes, so there are 4^2 = 16 possible zygotes.
- When n = 3, there are 2^3 = 8 possible male gametes and 8 possible female gametes, so there are 8^2 = 64 possible zygotes.
- By analogy, when n = 23, with 8,388,608 possible male gametes and 8,388,608 possible female gametes, there are $8,388,608^2$ = 7.03687×10^{13} possible zygotes. This represents huge human genetic diversity, even without the added potential for variation introduced by genetic crossing over, mutation and environmental influence. No wonder no one has a double.

In summary, with n pairs of chromosomes, there are $(2n)^2$ possible gametes.

> **» Pointer**
> Maternal and paternal chromosomes reassort to make gametes and recombine to make zygotes. With n pairs of chromosomes, there are 2n possible gametes and $(2n)^2$ possible zygotes.

1.4.2 Powers of 10

The power shows how many 0s there are after the 1, so 10^1 = 10, 10^2 = 100, etc.

We often use powers of 10 in Biology when we want to write large numbers but do not want to have lots of 0s. An example is the number of bacteria in a suspension. There may be 1 000 000 000 000 bacteria in 1 cm^3 of suspension, but it is far more convenient to write 10^{12}. When dealing with numbers over several orders of magnitude, indices or logarithms are often used.

Another use of indices in Biology is when you are dealing with very small numbers and want to make them easier to visualise and use. Imagine an experiment in which you are timing how long it takes for an indicator to change colour at different values of pH. Your measurements will be in seconds, and to calculate the rate, you use the expression rate = 1/time.

You then would plot a graph of your results with the pH on the x axis and the rate on the y axis. But the numbers for the rate will be very small. To make them conceptually easier to deal with, you could multiply them, for example, by 100. You would, however, need to indicate that this is what you have done, by including '× 10^2' in the column heading. Here is a table of results that shows how to do this:

> **» Pointer**
> A decimal number with one digit before the decimal point is described as standard notation. It is a very useful way of expressing numerical results when you do your practical work.

much easier to plot

pH	time for indicator to change colour / s	reaction rate = 1/time / s^{-1} (4dp)	reaction rate × 10^2 / s^{-1} × 10^2 (2dp)
3	120	0.0083	0.83
5	80	0.0125	1.25
7	40	0.0250	2.50
9	90	0.0111	1.11
11	140	0.0071	0.71

from same reading

1.4.3 Negative indices

Another way of writing $1/x$ is x^{-1}. Another way of writing $1/x^2$ is x^{-2}. This way of writing numbers is most useful in Biology when we consider powers of 10. For example, you might want a suspension of bacteria to be one tenth of its original concentration, in other words a 1 in 10 dilution. To make it, you would take 1 cm³ of a bacterial suspension and add it to 9 cm³ water. Then your bacteria, instead of being in 1 cm³ are now in 10 cm³ so have been diluted to one tenth of their original concentration. You may see this written as a dilution of 10^{-1}.

1.4.4 Negative indices in units

You may see a rate written as minutes^{-1} or a concentration written as mol dm^{-3}. The negative index here means 'per', so the rate would be read as 'per minute' and the concentration as 'moles per dm³'.

1.4.5 Indices and arithmetic

a) Multiplication – if you wish to multiply 10 000 × 10 000 000, it is simplest to convert the numbers to standard form so the calculation becomes $10^4 \times 10^7$. The answer is found by adding the indices , i.e. $10^4 \times 10^7 = 10^{4+7} = 10^{11}$.

b) Division – if you wish to divide 1 000 000 by 100 000, it is simplest to convert the numbers to standard form so the calculation becomes $10^6 \div 10^5$. The answer is found by subtracting the indices, i.e. $10^6 \div 10^5 = 10^{6-5} = 10^1$.

» Pointer
In standard form, add the indices to multiply; subtract the indices to divide.

quickfire» 1.16

1 cm³ of bacterial suspension produces 250 colonies on an agar plate. Give the concentration of the original suspension in standard notation, if the suspension had been produced by three 1 in 10 serial dilutions.

Serial dilutions

If you want to make a range of dilutions of a bacterial suspension, it is more useful to have a sequence of dilutions that are ten times more dilute than the next in the sequence, rather than, for example, half. If you have a suspension, you may want to make dilutions that are a tenth, a hundredth and a thousandth of the original. Using powers of 10, this could be expressed as dilutions of 10^{-1}, 10^{-2} and 10^{-3}.

Take 1 cm³ of a bacterial suspension and add it to 9 cm³ water to give a 10^{-1} dilution, as described above. Then take 1 cm³ of the 10^{-1} dilution and add it to 9 cm³ water. Now the bacteria that were in 1 cm³ of the 10^{-1} dilution are now in 10 cm³ so instead of having $\frac{1}{10}$ of the original concentration, you have $\frac{1}{100}$, i.e. a dilution of 10^{-2}. 1 cm³ of 10^{-2} dilution added to 9 cm³ water gives 10^{-3} dilution.

This is called serial dilution because each new dilution is made from the previous one, in a series.

In this example, 1 cm³ is added to 9 cm³ water each time, giving a $\frac{1}{10}$ or 10^{-1} dilution with each step. In a similar way, a $\frac{1}{100}$ or 10^{-2} dilution could be made by adding 0.1 cm³ to 9.9 cm³ water each time.

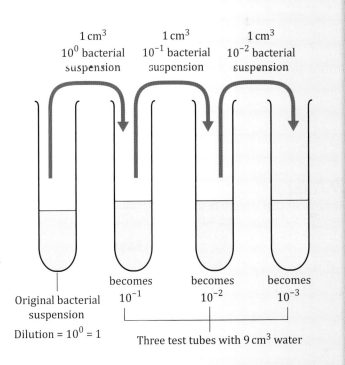

1 cm³ 10^0 bacterial suspension

1 cm³ 10^{-1} bacterial suspension

1 cm³ 10^{-2} bacterial suspension

Original bacterial suspension
Dilution = 10^0 = 1

becomes 10^{-1}

becomes 10^{-2}

becomes 10^{-3}

Three test tubes with 9 cm³ water

A bacterial suspension was serially diluted $\frac{1}{10}$ of its initial concentration 4 times. 2 cm³ of that dilution produced 34 colonies on a Petri dish. How many bacteria were in the initial, undiluted sample?

2 cm³ of diluted suspension gave 34 colonies, i.e. contained 34 bacteria

∴ 1 cm³ of diluted suspension contained $\dfrac{34}{2}$ = 17 bacteria.

The suspension had been diluted $\dfrac{1}{10} \times \dfrac{1}{10} \times \dfrac{1}{10} \times \dfrac{1}{10} = \dfrac{1}{10\ 000} = 10^{-4}$ times

∴ 1 cm³ of undiluted suspension contained $\dfrac{34}{2 \times 10^{-4}} = 17 \times 10^4$
= 1.7×10^5 bacteria

In summary: initial concentration $= \dfrac{\text{number of colonies}}{\text{volume of sample} \times \text{dilution}}$

1.4.6 Using logarithms

In Biology, we use logarithms to the base 10. The $\log_{10}$ of a number is the index or power of the number expressed to base 10, e.g. $\log_{10} 100 = \log_{10} 10^2 = 2$. In this example, the log is 2 and the antilog is 100.

Using the calculator to find logs and antilogs

To find a log using your calculator, press the log function, enter the number of which you need the log and press =

For example, to find $\log_{10} 1000$,

- press the log button
- press 1000
- press = to give the answer 3, because $1000 = 10^3$
  ∴ $\log_{10} 1000 = 3$

To find the antilog, press the 10^x button, enter the number for which you want to find the antilog and press =.

For example, to find the antilog of 5,

- press SHIFT log (to use 10^x)
- press 5
- press = to give the answer 100 000, because $\log_{10} 100\ 000 = \log_{10} 10^5 = 5$

A common use of logs in Biology is in the pH scale, where a tenfold increase in the concentration of H⁺ ions gives a decrease of 1 pH unit. Here is the equation: $pH = -\log_{10} [H^+]$.

» Pointer

Remember that square brackets indicate the molar concentration.

Example 1

$$[H^+] = 0.000\ 001 = 10^{-6}\,\text{mol dm}^{-3}$$

$$\log_{10} 10^{-6} = -6$$

$$-\log_{10} 10^{-6} = 6$$

∴ for $[H^+] = 0.000\ 001$, pH = 6

Example 2

$$[H^+] = 0.000\,01 = 10^{-5}\,mol\,dm^{-3} \text{ i.e. ten times more concentrated}$$

$$\log_{10} 10^{-5} = -5$$

$$-\log_{10} 10^{-5} = 5$$

$$\therefore \text{ for } [H^+] = 0.000\,01, pH = 5 \text{ i.e. one pH unit lower}$$

To find the $[H^+]$ from the pH, you work backwards:

If $pH = 4$, then $-\log_{10} [H^+] = 4 \therefore [H^+] = 10^{-4}\,mol\,dm^{-3} = 0.0001\,mol\,dm^{-3}$

Other examples of the use of logs in Biology are:

- Log scales on graphs, on pages 53–54
- Population growth curve of bacteria, on pages 71–72

1.5 Ratios

In biological systems, sometimes it is not the numbers of molecules, such as enzyme and substrate molecules, or the numbers of organisms, such as predators and prey, that matter, but their relative proportions. A useful way of showing this is with a ratio. When you express a ratio, it is always simplest to reduce it as far as possible. Then, instead of saying the ratio of rabbits to foxes in a given area is $300 : 3$, you would simplify the ratio to $100 : 1$.

>> Pointer

Always simplify a ratio as much as possible, e.g. $6 : 1$, not $12 : 2$.

In this next example, you are asked to calculate the ratio of the lengths of the head and body of a person at different stages of development.

You do not need to measure with a ruler, because there are grid lines on the diagram, which you can count.

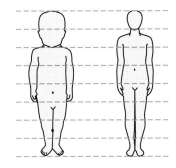

Always write the expression first:

Ratio = head length : body length

Then plug in the numbers:

For the adult, ratio = $1 : 8$

For the baby, ratio = $2 : 8$

2 and 8 are both divisible by 2 so the ratio can be simplified to $1 : 4$.

In a Biology exam, you may be asked to make a comment on the difference between these two ratios. You could discuss the observation that the large size of the baby's brain in proportion to its body size may be related to the cells of the brain at birth having more connections then than at any other time of life. These connections are progressively pruned as the brain continues its development until the end of adolescence. Other comments might relate to the observation that if the head were any larger, because of the dimensions of the pelvis and birth canal, it would be very difficult to give birth, so that natural selection may be limiting the maximum size of a baby's head. There are also evolutionary considerations. Perhaps, if the head were smaller, the brain would be less developed and it would take longer for the individual to become independent, so more parental input would be needed. In addition, if the head were smaller, it would have been a waste of energy for women to grow structures larger than needed, and therefore, there is a selective advantage in having the head and the female structures exactly matching in size.

quickfire 》》 1.17

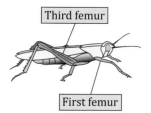

Third femur

First femur

The top joint of an insect's leg is its femur. Find the ratio of the lengths of the 1st femur : 3rd femur and suggest how this may be related to the way the locust moves.

When answering an examination question that requires you to calculate a ratio, make sure you use the correct figures and get the ratio the right way round. In the example below, in finding the biochemical oxygen demand (BOD) ratio between water containing raw sewage compared with water containing treated sewage, the 'compared with' comes second.

Water type	BOD / mg dm⁻³
Clean	2
Polluted	12
With raw sewage	360
With treated sewage	24

BOD ratio = water with raw sewage : water with treated sewage

= 360 : 24

= 15 : 1

Biologists often discuss the ratio of surface area : volume and ratios are very important when we consider the outcome of genetic crosses. These uses of ratios will be discussed later.

Respiratory quotient

When cells respire glucose aerobically, the volume of oxygen used and the volume of carbon dioxide produced are equal, as shown by the equation

$$C_6H_{12}O_6 + 6O_2 \longrightarrow 6CO_2 + 6H_2O$$

If other substrates are respired or an organism is carrying out both aerobic and anaerobic respiration, then the volumes are not the same. Other substrates must be oxidised first, so there is proportionally more oxygen used than with glucose only.

The ratio volume of carbon dioxide produced : volume of oxygen used is the respiratory quotient, RQ. It is written in equation form as

$$RQ = \frac{\text{volume of carbon dioxide evolved}}{\text{volume of oxygen used}}$$

》 Pointer

$$RQ = \frac{\text{volume of } CO_2 \text{ evolved}}{\text{volume of } O_2 \text{ used}}$$

The RQ can be calculated from a balanced chemical equation. Consider the respiration of glucose. The equation shows the number of molecules and, therefore, the volume of gases involved:

$$RQ = \frac{\text{volume of carbon dioxide evolved}}{\text{volume of oxygen used}} = \frac{6}{6} = 1$$

Consider the respiration of a fat, tripalmitin, $C_{51}H_{98}O_6$

$$2C_{51}H_{98}O_6 + 145O_2 \longrightarrow 102CO_2 + 98H_2O$$

$$RQ = \frac{\text{volume of carbon dioxide evolved}}{\text{volume of oxygen used}} = \frac{102}{145} = 0.7$$

This is the theoretical RQ when this fat alone is respired. So in an experiment, if an RQ were found to be approximately 0.7, you could assume that the main respiratory substrate is fat.

The volumes for aerobically respiring organisms are found using a respirometer. The diagram to the right shows the principle.

There are three parts to RQ calculations:

1   All the carbon dioxide produced is absorbed by the sodium hydroxide so any change in the position of the meniscus is because of the absorption of oxygen from the air in the boiling tube.

If the manometer tube radius r and height moved by meniscus is H, total volume of oxygen absorbed = volume meniscus has moved through = $\pi r^2 H$.

This is the volume of oxygen used. The oxygen is used in two ways:

a)   the respiration of glucose, which produces carbon dioxide

b)   the respiration of other substrates, which does not produce carbon dioxide.

2   To find how much oxygen is used oxidising subtrates other than glucose, the sodium hydroxide is removed. Oxygen used for glucose respiration will be equally balanced by carbon dioxide production so if the meniscus moves, it is because oxygen is used without carbon dioxide being produced. This represents the volume of oxygen used in substrate oxidation. If the meniscus moves through height h, the volume of oxygen used = $\pi r^2 h$.

So the volume actually used in the respiration of glucose

$$= \begin{array}{c} \text{total volume of} \\ \text{oxygen used} \end{array} - \begin{array}{c} \text{oxygen volume used in respiration} \\ \text{of substrates other than glucose} \end{array}$$

$$= \pi r^2 H - \pi r^2 h$$

3   The equation for respiration shows that the volume of carbon dioxide produced by glucose respiration is the same as the oxygen used, i.e. ($\pi r^2 H - \pi r^2 h$).

From 1, 2 and 3, RQ $= \dfrac{\text{volume of carbon dioxide evolved}}{\text{volume of oxygen used}} = \dfrac{\pi r^2 H - \pi r^2 h}{\pi r^2 H}$

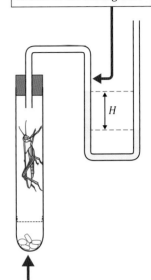

Liquid level in manometer goes up on this side as oxygen is absorbed from the air in the boiling tube

Sodium hydroxide pellets absorb all the carbon dioxide released so the volume change observed is only due to oxygen absorbed

---

Worms weighing 8 g in total are placed in a respirometer at 30 °C for 20 minutes and the meniscus level rises through 48 mm in a tube which is 1 mm diameter. In the absence of sodium hydroxide the meniscus moves a further 14 mm. Find the RQ for the maggots and suggest what substrate they may be respiring.

Tube radius $= \frac{1}{2} \times 1 = 0.5$ mm

Total volume of oxygen used $= \pi r^2 H = \pi \times 0.5 \times 0.5 \times 48 = 37.7$ mm^3

Volume of oxygen used in substrate oxidation only $= \pi r^2 h$

$= \pi \times 0.5 \times 0.5 \times 14 = 11.0$ mm^3

∴ volume of oxygen used in aerobic respiration $= 37.7 - 11.0 = 26.7$ mm^3

$= $ volume carbon dioxide produced

RQ $= \dfrac{\text{volume of carbon dioxide evolved}}{\text{volume of oxygen used}} = \dfrac{26.7}{37.7} = 0.7$

The theoretical value for RQ when protein is oxidised is between 0.5 and 0.8. The theoretical value for fat oxidation = 0.7. Of course more than one substrate may be oxidised at any one time, but the value of RQ = 0.7 calculated here suggests that fat is being oxidised.

1.6 Fractions

Once again, to repeat what you knew at GCSE, a fraction is a way of showing what is divided by what. So when we write $\frac{1}{4}$, we mean that we are considering a fourth part of something.

When we write $\frac{2}{8}$, both the numerator (at the top) and denominator (at the bottom) are divisible by 2, so this is the same as writing $\frac{1}{4}$, i.e. $\frac{2}{8} = \frac{1}{4}$.

If we write $\frac{3}{16}$, this is the same as writing $3 \times \frac{1}{16}$.

Remember that $10^{-2} = \frac{1}{100}$ so if you have 10^{-2} as a numerator, then you can rewrite the fraction, with 10^2 as the denominator:

$$\frac{400 \times 10^{-2}}{8} = \frac{400}{8 \times 10^2} = \frac{400}{800} = 0.5$$

If you have 10^{-2} in the denominator, that is the same as having 10^2 in the numerator, for example:

$$\frac{0.5}{25 \times 10^{-2}} = \frac{0.5 \times 10^2}{25} = \frac{0.5 \times 100}{25} = \frac{50}{25} = 2.0$$

We will come back to fractions later on when we use them in equations.

quickfire »» 1.19

a) 78.2×10
b) 56.475×1000
c) 0.02×10^2
d) 124.78×10^{-2}

» Pointer
Remember to write in every step of the calculation.

1.7 Decimals

1.7.1 Decimal notation

Decimal notation is standard in science. In the UK and North America, we show the decimal point as a full stop. In mainland Europe, however, a comma is used. But in the UK, a comma is a separator in numbers with thousands, such as when you write 3,000, meaning three thousand. In mainland Europe, 3,000 would mean three point nought nought nought instead. Take care to use the appropriate convention because your examiner will not try and guess which part of the world you are from.

quickfire »» 1.18

a) $\frac{1}{5} + \frac{3}{5}$
b) $\frac{2}{3} \times \frac{1}{2}$
c) $\frac{7}{8} - \frac{3}{8}$

Sadly, many students take out their calculator to multiply or divide by 10, 100, 1000, etc. It is not necessary. The beauty of decimals is that all you do is move the decimal point. If you are multiplying, it moves to the right, the same number of places as there are 0s in the number you are multiplying by. If you are dividing, the decimal point moves to the left, by the same number of places as there are noughts in the number you are dividing by:

multiplying by 10 so the decimal point moves one place to the right

| Multiplying: | 61.30×10 | = | 613.0 |
| Dividing: | $1226 \div 100$ | = | 12.26 |

dividing by 100 so the decimal point moves two places to the left

quickfire »» 1.20

a) $78.2 \div 10$
b) $56.475 \div 1000$
c) $0.02 \div 10^2$
d) $124.78 \div 10^{-2}$

1.7.2 The number of decimal places

When you give a decimal value, the number of figures after the decimal point is full of meaning. If, for example, you are weighing to find a dry biomass, 5 g is not the same as 5.0 g. The number of decimal places is an indication of the accuracy to which you are giving the value; 5 g means that the actual mass is between 4.5 g and 5.4 g and you are only accurate to 1 g. But 5.0 g means you are accurate to one decimal place, or 0.1 g, and the true value is between 4.95 g and 5.04 g.

It is important to use the correct number of decimal places to convey your level of accuracy. If you have a thermometer that is calibrated to 0.1 °C, you would give your reading to one decimal place. Then, it is clear that you are accurate to 0.1 °C, but no more or less accurate than that.

Stopwatches may give a reading of seconds with three decimal places. This means you can read them correctly to ± 0.001 s. But this degree of accuracy may not be appropriate if, for example, you are timing a colour change. In such a situation, you should only read to the nearest second. Colours may change slowly. In addition, the judgment of colour, the speed at which you make your decision, the time taken to operate the stopwatch and then have the stopwatch respond all take much longer than 0.001 s. In fact, that sequence of events takes just under a second and so if any decimal places are used, they cannot be correct.

If you make a calculation, in the answer you should only give the number of decimal places that are present in the raw data. Your answer cannot be more accurate than the figures used to calculate it:

Speckled peppered moth

You may remember *Biston betularia*, the peppered moth from GCSE. People have been studying it since the 1890s and the results of one release and recapture experiment are shown in the table. You are asked to calculate the percentage of each type recaptured. To do this, you write the fraction recaptured and then multiply by 100 to make a percentage:

Variety	Melanic	Non-melanic
Number marked and released	154	64
Number recaptured	82	16
Fraction recaptured	$\dfrac{82}{154}$	$\dfrac{16}{64}$
Percentage recaptured	$\dfrac{82}{154} \times 100 = 53$	$\dfrac{16}{64} \times 100 = 25$

In calculating the percentage of melanic moths recaptured, the calculator may give an answer to six or more decimal places, depending on the model. But the data are to 0 decimal places, because you can't have a fraction of a moth, which means you are not justified in quoting an answer with decimal places.

Students are often tempted to quote all the decimal places their calculator gives. Their answer is not more accurate. It is actually less accurate, as you cannot be more accurate than the initial data.

1.7.3 Rounding up and down

Your calculation may give an answer to two decimal places but if the raw data are all to one decimal place, then you must round your answer up or down. The rules are simple.

If your second decimal place is between 0 and 4, you round down. Here are some examples:

Number to 2 dp	Number to 1 dp
3.00	3.0
3.01	3.0
5.62	5.6
8.73	8.7
12.94	12.9

But if the number in the second decimal is 5 or above, then you round up:

Number to 2 dp	Number to 1 dp
14.65	14.7
14.66	14.7
23.87	23.9
32.18	32.2
43.99	44.0

If you have a number to several decimal places but you need to round it to one, the same principles apply:

Number to many dp	Number to 1 dp
3.234	3.2
5.427	5.4
7.68234	7.7

The only unusual situation happens when you have 44X after a decimal point, where X is greater than 4.

Consider rounding the number 3.442 to one decimal place. It can be done in two steps:

i) rounding 3.442 to 2 dp gives 3.44 because the 2 in the third decimal place is less than 5.

ii) rounding 3.44 to 1 dp gives 3.4 because the 4 in the second decimal place is less than 5.

But rounding 3.447 to one decimal place is different. Here are the two steps:

i) rounding 3.447 to 2 dp gives 3.45 because the 7 in the third decimal place is greater than 5.

ii) rounding 3.45 to 1 dp should give 3.5 because the 5 in the second decimal place is not less than 5. But it doesn't. This is because the original number, 3.447, is less than 3.5, so the first decimal place is not rounded up, even though, if you considered the decimal places individually, it would be.

It is conventional to write in how many decimal places you have corrected to so if you were quoting the temperature of a water bath, you might write 75.6 °C (1dp) or the mass of cells produced is 18.654g (3dp).

≫ Pointer
Don't forget to round your answer so you have the appropriate number of decimal places.

1.7.4 Significant figures

Significant means having meaning, so significant figures are the figures that have meaning.

1 If you have a zero at the beginning of a number, e.g. 08, the 0 has no meaning and is not a significant figure.

2 But ending a figure with a 0 is significant, after all 20 does not mean the same as 2.

3 If the zeros follow a decimal point, they are significant, e.g. 0.5 does not mean the same as 0.05, so the zeros following the decimal point are significant.

4 But at the end of the number after the decimal point they are not, e.g. 0.05 has the same value as 0.050, so the final zero is not significant. This, of course, is not the case when there is a unit following the number, as discussed on page 25, but it is true for pure numbers.

5 Depending on which calculator you use, you may get answers with 12 or more digits. As with the comment on decimal places above, the answer cannot be more precise than the data that was entered into the calculation. If the input data had, say, 2 significant figures, the output should not have more than two.

≫ Pointer
Make sure you have the appropriate number of significant figures.

1.8 Negative numbers

Negative numbers as used in Biology have very specific biological meanings. When considering the development of structures like the fingers and toes in a human foetus, you may come across the concept of negative growth. Human limb buds form in the fourth week of gestation. At the end of each limb bud, apoptosis, or programmed cell death, occurs so that four areas of cells fail to divide, leaving five digits, as indicated in the diagram below.

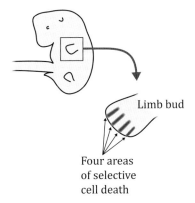

Limb bud

Four areas of selective cell death

Overall the hand or foot is growing and getting larger, but its development requires that four areas do not and so they show negative growth. Increases in cell mass or number for these areas would then be described by negative numbers.

Water relations in plant cells

A common use of a negative number is in quantifying the osmotic or solute potential of a solution and the water potential of a cell. The more concentrated a solution, the more the pull on water molecules to move into it. The water potential of a cell can be thought of as the pull of water into the cell. Its maximum value is 0, which is the water potential of pure water. The more negative the water potential, the stronger the pull, so water always moves by osmosis towards the most negative water potential. A pull inwards has a negative value and the stronger the pull, the more negative its value.

A push, such as a pressure potential, with the cell contents pushing outwards on the cell wall, would have a positive value. Imagine two adjacent plant cells, as shown below:

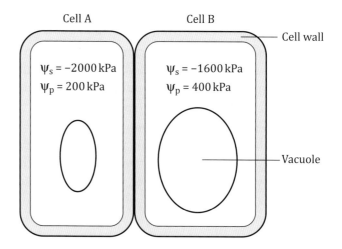

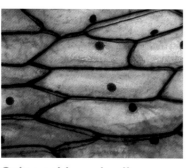

Pointer

The more negative the water potential the greater the pull on water.

quickfire》》 1.22

Find the water potential of a cell which has a solute potential of –1500 kPa and a pressure potential of 500 kPa.

quickfire》》 1.23

Find the pressure potential of a cell with a solute potential of –1500 kPa and a water potential of –2000 kPa.

Pointer

Water always moves towards the more negative water potential.

As you were reminded on page 10, the water potential of the cell, i.e. the balance between the two opposing forces, indicates the actual direction of water movement. The water potential for each cell can be calculated, as shown here:

First give the word equation:

$$\text{water potential} = \text{pressure potential} + \text{solute potential}$$

Then give the symbols: $\psi_c = \psi_p + \psi_s$

For cell A: $\psi_c = \psi_p + \psi_s$
$$= 200 + (-2000)$$
$$= -1800 \text{ kPa}$$

For cell B: $\psi_c = \psi_p + \psi_s$
$$= 400 + (-1600)$$
$$= -1200 \text{ kPa}$$

Cell A has a more negative water potential, so water will flow into it from cell B. As more water flows into cell A, its vacuole will get bigger and push more on the cell wall, so its pressure potential will become more positive. (Let's say it goes up from 200 kPa to 300 kPa.) At the same time, the contents of cell A are becoming more dilute. But the numerical value of solute potential remains unchanged.

Only very small volumes of water move and although the concentrations change slightly, the solute potentials change much less because the scale is not linear but logarithmic. This means that the concentration inside the cell would have to change 10-fold to make a difference to the solute potential of 1 kPa.

The new water potential is $\psi_P + \psi_s = 300 + (-2000) = 300 - 2000 = -1700$ kPa. Previously the water potential was -1800 kPa, so at -1700 kPa, it is pulling less on the water than it was before.

At the same time cell B is losing water so its pressure potential will go down. (Let's say it goes down from 400 kPa to 300 kPa.) Its solute potential is unchanged.

The new water potential is $\psi_P + \psi_s = 300 + (-1600) = -1300$ kPa. Previously the water potential was -1200 kPa so at -1300 kPa, cell B is going to be pulling more on the water than it was before, but it is still pulling less on the water than cell A.

The water movement continues until cell A and cell B are pulling on the water equally. When that happens there is no net movement of water and the two cells are in equilibrium. The new water potential then is the average of the two original water potentials.

quickfire 》 1.24

What is the water potential of two cells at equilibrium if their initial water potentials are -2000 kPa and -1200 kPa?

$$\text{Water potential at equilibrium} = \frac{\text{initial } \psi_c \text{ for cell A} + \text{initial } \psi_c \text{ for cell B}}{2}$$

$$= \frac{-1800 + (-1200)}{2}$$

$$= \frac{-3000}{2}$$

$$= -1500 \text{ kPa}$$

——— Don't forget the units

The new pressure potential

As cell B now has less water, its pressure potential will be less and as cell A has more, its pressure potential will be more. It is possible to calculate the new pressure potentials at equilibrium. The new water potential is the mean of the initial values (-1500 kPa). The solute potential is unchanged.

》 Pointer
Sometimes pressure potential is called 'turgor pressure' and solute potential is called 'osmotic potential'. Check your specification to see which terms it uses.

For cell A:	For cell B:
new ψ_c = new $\psi_p + \psi_s$	new ψ_c = new $\psi_p + \psi_s$
$\therefore$ new ψ_p = new $\psi_c - \psi_s$	$\therefore$ new ψ_p = new $\psi_c - \psi_s$
$= -1500 - (-2000)$	$= -1500 - (-1600)$
$= -1500 + 2000$	$= -1500 + 1600$
$= +500$ kPa	$= +100$ kPa

This result is sensible. We expect cell A to have a higher pressure potential than at the start as it has gained water. We expect cell B to have a lower pressure potential than at the start as it has lost water.

Test yourself 1

1 If 34% of the bases in a molecule of DNA are thymine, what percentage of the bases are guanine?

2 At maximum inspiration, the air pressure in the alveoli is 0.30 kPa below atmospheric pressure.

At maximum expiration, the alveolar pressure is 0.29 kPa above atmospheric pressure. Calculate the difference in alveolar pressure during one cycle of breathing.

3 One complete cardiac cycle lasted from 0.50 seconds to 1.34 seconds after measuring began. Use this information to calculate the heart rate.

4 During exercise, the coronary arteries carry 1200 cm^3 blood to the heart muscle each minute. At rest they carry 300 cm^3. What is the ratio of the rate of blood flow to the heart muscle during exercise to the rate at rest?

5 The data in the table show the progress of fungus disease in potato plants. One group has been sprayed with a copper-based fungicide and the other has not.

Day	Amount of infection / AU	
	Unsprayed	Sprayed
5	8.5	7.0
10	15.5	7.7

a) Find the rate of infection in unsprayed between days 5 and 10.

b) Calculate the ratio of infection rate of sprayed : unsprayed plants between days 5 and 10.

6 The table below shows the absorbance of a suspension of cells of a green alga, as measured in a colorimeter. Calculate the mean daily rate of absorbance increase between days 1 and 4.

Time / days	Absorbance / AU
0	0.00
1	0.06
2	0.43
3	0.63
4	0.78

7 The data in the table indicate glucose uptake in ileum tissue over a period of 45 minutes, in the presence or absence of sodium ions. Estimate the relative effects on uptake with and without sodium.

Time / minutes	Glucose uptake / μmol /g ileum tissue	
	with sodium ions	without sodium ions
0	0	0
5	70	3
15	210	8
25	360	15
35	500	23
45	630	33

⑧ It is estimated that a hectare of clover plants contains 109.6 kg nitrogen, of which 96.0 kg is derived by fixation by *Rhizobium* in their root nodules. How much nitrogen do they obtain from their only other nitrogen source, inorganic ions in the soil?

⑨ A 0.5 m × 0.5 m square gridded quadrat was used to count the number of wood sage plants in a coppiced woodland. Each cell of the grid is 10 cm × 10 cm. The number in each cell is indicated below. What is the density, i.e. mean number of plants per m²?

3	4	2	2	0
2	4	6	3	1
2	5	2	0	2
1	3	3	0	0
2	4	7	5	3

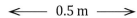

←—— 0.5 m ——→

⑩ Here is a sequence of bases in an mRNA molecule:

G U C C U A C G U A C C C U G A A A

How many amino acids does the sequence code for?

⑪ Complete the table to show the percentage of guanine in cow DNA and the percentage of adenine, guanine and thymine in yeast DNA.

Organism	Percentage of each base in DNA			
	thymine	cytosine	adenine	guanine
cow	29	21	29	
yeast		19		

⑫ The growth of duckweed, *Lemna minor*, was investigated by placing 30 plants in a pond and counting how many plants there are for the next 24 days. From the counts in the table below, calculate the average population growth rate for the first 12 days of the experiment.

Time / days	Number of plants
0	30
4	90
8	190
12	310
16	510
20	690
24	780

↑ 0.5 m ↓

⑬ The protein albumen was incubated with a protease for 3 hours at 45 °C. The average rate of production of amino acids was 95 mg dm⁻³ h⁻¹. The experiment was continued for a further 5 hours, in which an additional 246 mg amino acids were produced. Calculate the mean rate of amino acid production over the whole 8 hours.

⑭ The complete hydrolysis of glucose can produce 38 molecules of ATP. ATP can provide approximately 30.7 kJ mol⁻¹ usable energy. Calculate the energy capture by ATP in the aerobic breakdown of 1 mole of glucose.

Chapter 2

Processed numbers

In Biology, you will often have numbers that need to be processed so that you can make biological sense of them. This section will show you how to decode the question asked so that you can make the correct calculation. It is generally the use of language, rather than the maths, that is the problem.

2.1 Percentage

One very common operation is finding a percentage, but students often fail to make the appropriate calculation because they have not understood what is being asked.

2.1.1 Per cent calculation

'Per cent' means 'by 100' so when you calculate a per cent, it is as if you were finding the number out of 100.

When faced with a problem, if you're not sure how to proceed, give yourself an easy example first so you can see what to do. Supposing you are asked to find the percentage of black-tailed godwits in their summer plumage, given that 288 of these birds have their summer colours, out of a total population of 720. You may not even have heard of godwits, but you do not need to know that these are wading birds found in wet meadows and grassy marshes, to work out the answer.

First, make up a calculation based on easy numbers. Pretend that 3 out of a population of 6 have their summer coloration. You know intuitively, that 3 out of 6 is 50% because you know that 3 is half of 6. Now rearrange the numbers 3, 6 and 100 to make an equation that gives you 50% as your answer. Because these numbers are easy, you can see $\frac{3}{6} \times 100 = 50\%$.

So this is how you work out the problem. Substitute the numbers given to you to find the answer:

A godwit

» Pointer

$\% = \dfrac{\text{number given}}{\text{total number}} \times 100$

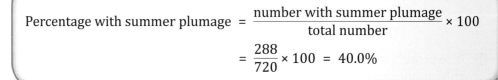

$$\text{Percentage with summer plumage} = \frac{\text{number with summer plumage}}{\text{total number}} \times 100$$

$$= \frac{288}{720} \times 100 = 40.0\%$$

2.1.2 Per cent increase and decrease

A change in mass can be expressed as a percentage change. If mass has been gained, the percentage change is positive and if mass has been lost, the percentage change is negative. To find a percentage change, you must first find the actual change. Then you work it out as a percentage of what you started with.

Here are examples of the change of mass of different desert rodents in response to the same diet of dry food with no supplementary water, as is often the case in their habitats. *Gerbillus* increases its mass from 35.0 g to 37.1 g but the mass of *Acomys* decreases from 37.4 g to 27.7 g.

quickfire 2.2

Raw cabbage contains 49 mg vitamin C per 100 g. Boiled cabbage contains 20 mg vitamin C per 100 g. Calculate the percentage loss of mass.

You may be asked to find the per cent change in mass.

For *Gerbillus*,

$$\% \text{ mass increase} = \frac{\text{actual mass increase}}{\text{intial mass}} \times 100 = \frac{37.1 - 35.0}{35.0} \times 100 = \frac{2.1}{35.0} \times 100 = +6\%$$

For *Acomys*,

$$\% \text{ mass increase} = \frac{\text{actual mass increase}}{\text{intial mass}} \times 100 = \frac{37.4 - 27.7}{37.4} \times 100 = \frac{9.7}{37.4} \times 100 = -25.9\%$$

For *Acomys*, the change is negative, because it lost mass.

But it is important to be sure of the wording of the question:

If you were asked, instead, to find, not the per cent change, but the per cent increase in mass for *Gerbillus* and the per cent decrease for *Acomys*, it would be the same for *Gerbillus* because unless a figure is qualified, it is assumed to be positive, so 6% means the same as +6%, in other words, an increase. For *Acomys*, however, you would not need the minus sign, because the question has asked for the decrease, not the change.

For *Gerbillus*,

$$\% \text{ mass increase} = \frac{\text{actual mass increase}}{\text{intial mass}} \times 100 = \frac{37.1 - 35.0}{35.0} \times 100 = \frac{2.1}{35.0} \times 100 = 6\%$$

For *Acomys*,

$$\% \text{ mass decrease} = \frac{\text{actual mass decrease}}{\text{intial mass}} \times 100 = \frac{37.4 - 27.7}{37.4} \times 100 = \frac{9.7}{37.4} \times 100 = 25.9\%$$

The per cent increase for *Gerbillus* is 6.0% but the percentage change can be written +6.0%. The per cent decrease for *Acomys* is 25.9% but the percentage change is −25.9%. Notice how the words are used. For *Gerbillus*, the increase is positive and the change is positive as it has gained mass. But for *Acomys*, the decrease is positive but the change is negative because it has lost mass. It is important that you pay attention to the words so that you express your answer properly.

As this is Biology, you could be asked about the meaning of the results, such as to suggest which rodent is better adapted to desert life. If you picture the organism and think about its niche, you are more likely to produce an answer that is biologically sensible. The answer would be *Gerbillus* because it can increase its mass even without water being available, which is presumably a good thing in the desert, whereas *Acomys,* on the same diet, decreases its mass. You can deduce that *Acomys* is not thriving so well and is therefore less well adapted.

》 Pointer

Mass gain is a positive mass change; mass loss is a negative mass change.

Gerbillus

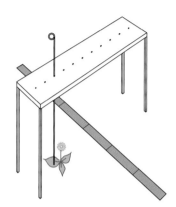

Point frame quadrat

2.1.3 Per cent frequency

Environmetal gradients can be assessed by measuring the per cent frequency of a plant along a transect. This can be done in two ways:

a) Point frame quadrat: this is a metal bar 0.5 m long with 10 holes, 10 cm apart. It is supported just above the transect, at intervals along it, at 90° to the line. At each distance along the transect, a needle is slotted vertically down through each of the 10 holes on to the plants below. If the needle touches the species in question, the species is scored. The number of times out of 10 a species is touched allows an easy calculation of percentage frequency at each position along the transect. The percentage frequency is the frequency of the plant species occurring out of 10 samples. If a daisy is touched with the needle in 4 positions at, say, 10 m along the transect, then we can say that the percentage frequency of daisies at 10 m is $\frac{4}{10} \times 100 = 40\%$.

b) Gridded quadrat: a square frame quadrat may be divided into a 10 × 10 grid, giving 100 squares. The quadrat is laid down along the transect and at each position, the numer of squares in which the plant species occurs is counted. The number of squares containing the plant, out of 100, gives the percentage frequency. It can be done with a 5 × 5 grid, but then, as there are only 25 squares, each represents $\frac{100}{25} = 4\%$ of the total. The number of squares containing the plant is multiplied by 4 to give the percentage frequency. Smaller squares give a more accurate estimate.

2.1.4 Per cent area cover

Quadrats can be used to estimate the percentage area covered by a plant species, and this is discussed on pages 75–76.

2.1.5 Percentage error

An error, in this context, does not mean a mistake. It refers to the uncertainty in a measurement related to the size of the calibration on the measuring equipment. Such an error occurs for every reading in an experiment and so it is described as a systematic error.

If, for example, the smallest graduations on a ruler are at 1 mm intervals, when a length is measured, there is an uncertainty at each end of the scale of up to 0.5 mm. The measurement of length can, therefore, only be accurate to ±1 mm, i.e. 0.5 mm at each end. The percentage error is affected by:

- The size of the graduation: a smaller graduation gives more accurate reading and reduces uncertainty.

- The size of the measurement: the uncertainty is proportionally smaller with a larger reading than with a smaller reading.

The percentage error, or uncertainty, can be calculated:

Example 1

$8.6\ cm^3$ water was measured with a $10\ cm^3$ syringe, calibrated to $1\ cm^3$.

$$\% \text{ error} = \frac{\text{accuracy}}{\text{measured length}} \times 100 = \frac{1}{8.6} \times 100 = 11.6\ \% \text{ (1 dp)}$$

Example 2

8.6 cm³ water was measured with a 50 cm³ burette, calibrated to 0.2 cm³.

$$\% \text{ error} = \frac{\text{accuracy}}{\text{measured length}} \times 100 = \frac{0.2}{8.6} \times 100 = 2.3 \% \text{ (1 dp)}$$

This example shows that measuring the same volume in a piece of equipment with smaller graduations gives a smaller percentage error.

Combining errors

If a calculation depends on several readings with inherent uncertainties, the result cannot be more accurate than the largest error allows. The error in the result depends on the sum of the individual errors.

Example

16.0 cm³ oxygen, collected by displacement of water in a burette, graduated to 0.2 cm³, are collected in 10 minutes, measured by a stop watch, correct to ±0.01 s.

$$\% \text{ error in timing} = \frac{0.01}{10} \times 100 = 0.1\%$$

Total percentage error
= 1.3 + 0.1 = 1.4%

$$\% \text{ error in volume} = \frac{0.02}{16.0} \times 100 = 1.3\%$$

$$\text{Rate of oxygen production} = \frac{16.0}{10.0} = 1.6 \text{ cm}^3 \text{ min}^{-1}$$

$$\text{Total actual error} = 1.4\% \times 1.6 = \frac{1.4}{100} \times 1.6 = 0.02$$

$$\therefore \text{ Rate of oxygen production} = 1.6 \pm 0.02 \text{ cm}^3 \text{ min}^{-1}$$

In general, the result and the error are quoted to the same number of decimal places but where the first number following the decimal point is small, as shown here, then an extra decimal place is given.

All the terms in a calculation should have the same unit. In calculating percentage errors, it is the errors that are important, not the factor being measured. So errors in time and volume, for example, may be combined.

2.2 Proportion

2.2.1 Surface area and volume

Why are bacteria so small? Could an elephant be the size of a mouse? Could a spider be the size of a fox? Why did dragonflies in the Carboniferous period have a wingspan of 800 mm, but today only 80 mm? Biologists have asked such questions many times and the more that we understand about physics and chemistry, the more we understand that living organisms are the right size both for their structure and for the way their bodies function in the environment in which they evolved.

A simple organism uses diffusion across its surface to absorb the food and oxygen it needs and to remove its waste. Gas molecules diffuse efficiently across the average distance between collisions, after which they may be sent in different directions. Consequently, if a molecule is required to travel a distance greater than the distance between collisions, its movement will not be efficient. Collisions will keep changing its direction. Organisms using diffusion must either transport over very small distances only or have a large surface area to compensate for the inefficiency. Single cells rely on diffusion for absorption and transport but mammals need the large surface areas of lungs and the ileum to provide enough area that the diffusion across them is adequate. They also need a transport system to move molecules around the body.

An aspect of this that may be tested in Biology courses examines the surface area : volume ratio of cells or whole organisms where they may be simplified to a geometric shape, such as a cube, a sphere or a cylinder. You may be expected to calculate areas, volumes or their ratios. You would not be expected to remember the equations for the surface area or volume of a sphere.

Organisms as cubes

You may be asked to do similar calculations thinking of organisms as cubes. In this case, you would be expected to remember how to calculate the area and volume. The table below shows you how to do the calculation for two imaginary animals, one with sides of 1 unit and one with sides of 2 units. It doesn't matter what the units are for this argument, because you are finding a ratio. It might help to think about approximately cubic organisms, like a toad or a rhinoceros, as long as you ignore the limbs.

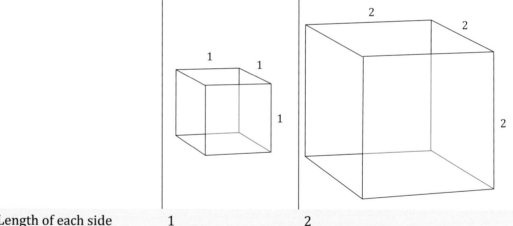

Length of each side	1	2
Area of each face	1 × 1 = 1	2 × 2 = 4
Number of faces	6	6
Total area	6 × 1 = 6	6 × 4 = 24
Volume	1 × 1 × 1 = 1	2 × 2 × 2 = 8
Surface area : volume ratio	6 : 1	24 : 8 = 3 : 1

The biological meaning of this is that for an organism of one unit length, for each unit of volume, there are 6 units of area. But if the organism has 2 units length for its dimensions, then each unit of volume has only 3 units of area. In other words, the bigger the organism, the less surface area it has for each unit of volume. This means that the bigger it is, the less likely diffusion alone can provide all the food and oxygen needed and remove all the carbon dioxide.

≫ Pointer
Diffusion is only efficient over small distances.

≫ Pointer
Organisms can be thought of as geometric shapes such as cubes, spheres and cylinders, in order to make calculations about their area and volume.

quickfire ≫ 2.4
Calculate the surface area : volume ratio of an organism simplified to a cube, with sides of 3 units.

In addition, the bigger the organism, the less area it has per unit volume through which it loses heat. The smaller organism has more area for each unit of its volume through which it will lose heat, which is consistent with the observation that of closely related species, those found in Arctic environments are larger than those found in temperate environments. Polar bears are bigger than the European brown bear and arctic foxes are longer than red foxes, supporting this argument. Similarly, babies have a much bigger surface area : volume ratio than adults and so they lose heat much more quickly, which is why they are more at risk of hypothermia than adults.

The table below shows how these figures increase as the size of the cubic organism increases:

Side length / units	Area of each side	Total surface area	Volume	Surface area : volume
1	1	6	1	6.0 : 1
2	4	24	8	3.0 : 1
3	9	54	27	2.0 : 1
4	16	96	64	1.5 : 1
5	25	150	125	1.2 : 1
6	36	216	216	1.0 : 1
7	49	294	343	0.9 : 1
8	64	384	512	0.8 : 1
9	81	486	729	0.7 : 1
10	100	600	1000	0.6 : 1

》Pointer
Think about what the numbers mean, in biological terms.

Now look at these three graphs:

1 Surface area and length of side

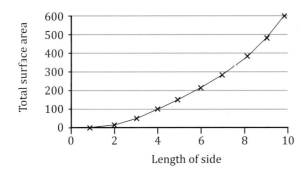

This graph shows that as the lengths of the sides increase, the surface area increases. The line is not straight. With longer sides, the area increases more and more.

2 Volume and length of side

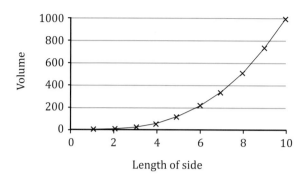

Length of side

This graph shows how volume increases as the length of the sides increases. As with the length, the increase is gradual with short lengths but becomes steeper as the length increases.

3 Surface area : volume ratio and length of side

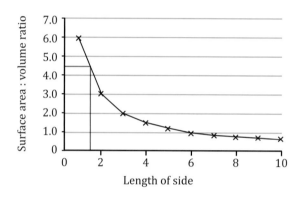

Length of side

>> *Pointer*

Surface area : volume ratio gets smaller as organisms get bigger.

quickfire >> 2.5

a) Estimate from the graph the surface area : volume ratio of a cube of side length 2.5.

b) Check your estimate by calculating the correct answer.

>> *Pointer*

Always draw on the graph, rather than estimating by eye. You are more likely to be accurate.

This graph shows how the ratio of surface area : volume changes as the length of the sides gets longer. The ratio goes down, as could be seen from the calculations for cubes of sides 1 and 2 above. As the sides get longer, the ratio decreases less and less, so that above about 10 units, the ratio does not change very much.

This graph says that the bigger the organism, the less surface area it has for every unit of its volume. But above a certain size, increase in size does not make much difference to the surface area : volume ratio. There are presumably other factors that limit the organism's size, such as limb strength, speed of nervous impulse or ability to produce blood pressure that is high enough.

A graph like this allows you to estimate the surface area : volume ratio for any length side, without having to calculate. For example, if you were asked to use the graph to estimate the surface area : volume ratio for an organism with sides of 1.5 units, drawing on the graph as shown, gives a ratio of 4.4 : 1.

You might be asked to relate the ratios for these imaginary animals to their habitats. You would be expected to explain that those that have a larger surface area : volume ratio are better adapted for absorbing materials, such as glucose, minerals or oxygen, across their surface and better adapted for releasing materials such as carbon dioxide or ammonia across the surface. You might suggest that these smaller organisms could be aquatic, because if they rely purely on diffusion, the greater efficiency of movement in water than in air makes them more likely to be adapted to living in water.

Organisms as spheres

A spherical bacterium has a diameter of 8 μm. Its area is found by the equation area = $4\pi r^2$ and its volume by the equation volume = $\frac{4}{3}\pi r^3$. Find the surface area : volume ratio for this bacterium.

The calculations are straightforward. The diameter will provide you with the radius, and the answer is found by substituting into the equations:

First the surface area: Write the equation	area = $4\pi r^2$
Find the radius	radius = $\frac{8}{2}$ = 4 μm
Substitute into the equation	area = $4 \times \pi \times 4 \times 4$ = 201 μm^2
Now find the volume: Write the equation	volume = $\frac{4}{3}\pi r^3$
Substitute into the equation	volume = $\frac{4}{3} \times \pi \times 4 \times 4 \times 4$ = 268 μm^2
Now find the ratio of area : volume	ratio = 201 : 268
Simplify the ratio by dividing both sides by 268 so that the answer comes out as x : 1	ratio = 0.75 : 1

A biological note: when ratios are simplified, it is normal to divide both sides by the smaller number. When biologists study surface area : volume ratio, however, they consider unit volume, so that it is easier to compare the area different structures have in relation to their volumes. You always make the volume = 1 and so, as in the example above, you divide both sides by the numerical value for the volume.

≫ Pointer
Write the equation in full to start with.

≫ Pointer
Simplify the ratio to give x : 1.

Organisms as cylinders

The third shape you might encounter is a cylinder. This could represent a worm.

On the right is a cylinder of radius r and length l.

The circumference of the cylinder is the circumference of a circle: $C = 2\pi r$

If you know the diameter of the cylinder, d, you can find the radius because it is half of the diameter:

$$d = 2r$$

To work out the area of the cylinder we can imagine it uncurled into a rectangle. The short side is the circumference of the circle and the long side is the length l.

Area of rectangle = length of short side × length of long side
= $2\pi rl$

Volume of a cylinder = area of circle × length
= $\pi r^2 \times l$

Surface area : volume = $2\pi rl : \pi r^2 l$

Divide both sides by $\pi r^2 l$

so ratio = $\frac{2}{r}$: 1

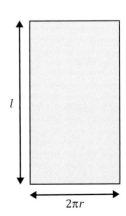

a) Calculate the surface area : volume ratio for two cylinders of the same length and of diameter 2 mm and 4 mm.

b) If these represented worms in the same habitat, which is likely to be more active.

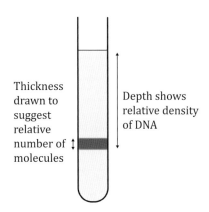

Thickness drawn to suggest relative number of molecules

Depth shows relative density of DNA

As the radius of the worm gets bigger the fraction $\frac{2}{r}$ gets smaller, so with increasing diameter, worms have a decreasing surface area : volume ratio. As a worm gets fatter, there will be a particular radius above which it will be too fat for diffusion to be efficient at allowing it to absorb or remove all it needs. To be fatter, the worm will need to have another mechanism for getting molecules and ions around, such as a blood circulatory system for transporting what has been absorbed, which is what the earthworm does. It still has to absorb oxygen across its surface, so the area puts a limit on how much can be absorbed. This is why there are no giant worms. In the Carboniferous and Permian periods, when the oxygen concentration in the atmosphere was higher, oxygen diffusion would have been more efficient. This suggests that worms may have been bigger then.

2.2.2 DNA replication

Another example of a situation where proportions are significant is in analysing DNA samples to demonstrate their mode of replication. DNA replicates semi-conservatively, which means that in each double helix, there is one original strand from its parent molecule and one newly synthesised daughter strand. If the parental, original molecules are labelled with heavy nitrogen, ^{15}N, they will sediment at a lower level than DNA containing light DNA, with ^{14}N. Any DNA molecule that contains one strand of each has an intermediate density and will sediment between the two. So the position of the band shows whether the DNA molecule has two heavy strands, two light strands or one of each.

The thickness of the bands as drawn in the diagrams suggests how much DNA is present. If one band has half the thickness of another, this indicates that it has half the number of molecules of DNA.

So from the depth of each band and the thickness as it is drawn in a diagram, it is possible to make deductions about relative amounts and composition of the DNA.

Parental generation:

Here is tube of sedimented DNA which contains parent molecules labelled with ^{15}N in both strands:

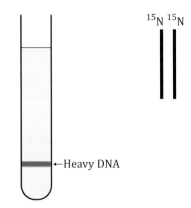

^{15}N ^{15}N

←Heavy DNA

Two strands labelled with heavy nitrogen.

1st generation:

When a molecule replicates, the two heavy strands separate and each synthesises a new daughter strand. If only nucleotides with light DNA are available, the complementary daughter strand will be light. So each molecule contains a heavy and a light strand and so is intermediate in density between completely heavy and completely light.

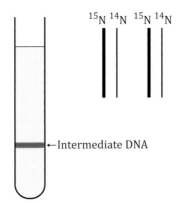

←Intermediate DNA

Two intermediate molecules each with two strands, one heavy and one light.

2nd generation:

When the first generation molecules replicate, the two strands separate and each synthesises a new complementary strand with nucleotides containing light DNA. So now, there are four molecules, two of which contain the original heavy parental strands and two which are entirely light. They are in equal proportions so the bands are of equal height.

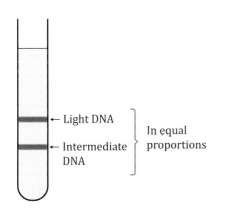

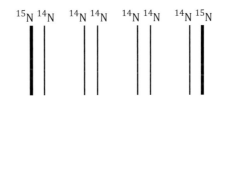

← Light DNA

← Intermediate DNA

} In equal proportions

≫ Pointer
The band thickness represents the proportion of molecules at that density.

Four molecules, two of which are intermediate and two of which are light.

3rd generation:

When these four molecules replicate, their strands separate and each synthesises a new light complementary strand. Two of the resulting eight molecules will each contain one of the original parental, heavy strands and one light strand, so they will be intermediate in density. Two molecules out of eight represents one quarter. So three quarters of the molecules will be light. The intermediate band is therefore drawn to be one third of the height of the light band.

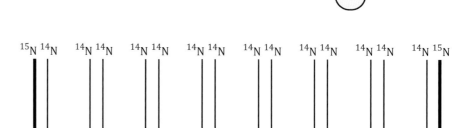

a) In the Meselson and Stahl experiment, how many molecules of DNA could be shown to have parental strands after 4 generations of replicating in medium containing ^{14}N?

b) What is the ratio of molecules containing parental strands : new molecules after five generations?

← Light DNA

← Intermediate DNA

^{15}N ^{14}N ^{14}N ^{14}N ^{14}N ^{14}N ^{14}N ^{14}N ^{14}N ^{14}N ^{14}N ^{14}N ^{14}N ^{14}N ^{14}N ^{15}N

Eight molecules, two of which are intermediate and six of which are light.

In summary:

Generation	Number of DNA molecules	Number of molecules containing parental strands	Number of molecules containing entirely new strands	Ratio of molecules containing parental strands : new molecules
Parental	1	1	0	1 : 0
1	2	2	0	1 : 0
2	4	2	2	1 : 1
3	8	2	6	1 : 3
4	16	2	14	1 : 7

2.3 The concentration of a solution

Knowing the concentration of a solution is important if you study physiology or biochemistry. It is crucial to know what units the concentration is given in. There are two common ways of describing concentration and giving the correct units is vital.

2.3.1 Per cent concentration

You may be given a solution of starch labelled '1% starch'. The % sign, as always, means 'out of 100' so this means there is 1 g of starch in 100 cm³ solution. You may mix this with a solution of 0.1% amylase, which means the enzyme solution has 0.1 g amylase in 100 cm³ solution.

When you make these solutions, all you have to do is weigh and measure. There are no calculations to be done. But do bear in mind when you make up 2% glucose, it is not 2 g glucose + 100 cm³ water. When you make a solution, the volume expands and so 2 g glucose + 100 cm³ water will have a volume greater than 100 cm^3. The way to do it is to dissolve the glucose in, say 70 cm^3 of water and when it is fully dissolved, make up the volume to 100 cm^3.

If you need 50 cm^3 4% salt solution, you may have to calculate how much salt you need. Always work things out from first principles. This is how:

1	Initial statement about solution	100 cm^3 4% salt solution has 4 g salt
2	Find how much in 1 cm^3	1 cm^3 4% salt solution has $\frac{4}{100}$ g salt
3	Find how much in the volume you want to make	50 cm^3 4% salt solution has $\frac{4}{100}$ × 50 g salt
4	Final answer	= 2 g salt

2.3.2 Using moles

For some reason, students are often frightened by moles. Perhaps they had a bad experience at GCSE. This section will show you how logical the calculations are and how you can do them easily.

The table on the right shows the atomic masses of some atoms common in biological molecules.

The relative molecular mass is the sum of the masses of all the atoms in a molecule. Often the word 'relative' is not used, because it is assumed that all atomic masses are relative to the mass of carbon, which is 12. Relative molecular mass has the symbol M_r. This table shows some calculations of M_r:

Name	Formula	Calculation	M_r
water	H_2O	1 + 1 + 16	18
carbon dioxide	CO_2	12 + 16 + 16	44
methane	CH_4	12 + (1 × 4)	16
urea	$CO(NH)_2$	12 + 16 + 2(14 + 1) = 12 + 16 + 30	58
glucose	$C_6H_{12}O_6$	(12 × 6) + (1 × 12) + (16 × 6) = 72 + 12 + 96	180

A mole is a mass. It is the relative molecular mass (M_r) given in grams. So if you have 18 g water or 180 g glucose, in each case, you have 1 mole of the substance.

2.3.3 Molarity

The concentration of a solution can be expressed in terms of the mass of solute in a given volume. If the mass were in grams, the concentration could be given as g/dm^3. If the mass were in moles it could be given as moles/dm^3. The number of moles in 1 dm^3 is the solution's molarity.

The M_r of glucose is 180 and one mole of glucose has a mass of 180 g. If 1 dm^3 of a solution contains 180 g glucose, the solution has a concentration of 1 mol/dm^3 or 1 mol dm^{-3}. In the past, the unit M, meaning 'molar' has been used, but the current convention is to use mol/dm^3 or mol dm^{-3}, where mol is the abbreviation for mole.

1 mmol is read as '1 millimole'. Milli- as a prefix means 10^{-3}. A solution of concentration 1 mmol dm^{-3} contains 10^{-3} moles in each dm^3.

quickfire 2.8

a) What is the mass of sucrose in 100 cm^3 of a 1% solution?
b) What is the mass of lead nitrate in 10 cm^3 of a 1% solution?
c) What is the mass of sodium chloride in 50 cm^3 of a 0.1% solution?
d) What is the mass of maltose in 200 cm^3 of a 2% solution?

Element	Atomic mass
C	12
H	1
O	16
N	14

quickfire 2.9

a) How many moles in 58 g urea?
b) How many moles in 32 g methane?
c) How many g in 2 moles carbon dioxide?

» Pointer

1 mol/dm^3 is the same as 1 mol dm^{-3}.

quickfire 2.10

Glucose has an M_r of 180. How many grams would you need in 1 dm^3 to make a solution that has a concentration of 1 mol dm^{-3}?

2.11

Fructose, an isomer of glucose, also has an M_r of 180. How many grams of fructose would you need in 1 dm³ to make a solution that has a concentration of 1 mmol dm⁻³?

Sometimes a Biology examination question will ask you about concentrations, but this is not a Chemistry test so all calculations will be very straightforward. For example, you may be asked how to make a 1 mol/dm³ solution of sucrose into a solution that is 0.5 mol/dm³:

A solution that is 0.5 mol/dm³ is half the concentration of a 1 mol/dm³ solution so you are being asked how to halve a concentration. If there are x moles in a given volume, you need to put them into double that volume. So you add an equal volume of water. Then you have x moles in twice the volume, so the concentration is halved. If you had 50 cm³ of a 1 mol/dm³ solution, you would add 50 cm³ water.

2.3.4 Hydrogen peroxide

Hydrogen peroxide is often used in biochemistry experiments to illustrate the properties of enzymes. It decomposes under the influence of the enzyme catalase, into oxygen and water:

$$2H_2O_2 \xrightarrow{\text{catalase}} 2H_2O + O_2$$

The concentration of hydrogen peroxide could be given as % or mol/dm³, as described above. However, the common way of expressing its concentration is to use the unit 'vol', which stands for volume, even though it is a measure of concentration. This is because its concentration is measured by the volume of oxygen it can generate. The higher its concentration, the more oxygen it can generate.

1 dm³ of a concentration of 1 vol generates 1 dm³ oxygen.

1 dm³ of a concentration of 10 vol generates 10 dm³ oxygen.

> **» Pointer**
>
> For hydrogen peroxide, vol is a unit of concentration.

> **» Pointer**
>
> Sometimes people write vols (with an s on the end) by mistake. Make sure you don't.

If you had a solution of 20 vol hydrogen peroxide and you wanted a solution of 5 vol, you would need to make it $\frac{1}{4}$ as concentrated. The number of molecules in a given volume would need to go into a volume four times as great. If you had 10 cm³ you would need to make it up to 40 cm³ by adding 30 cm³ water.

2.4 Biotic indices

A biotic index is a number calculated to describe a particular feature of a population. Different indices are suitable in different situations, so it is important which one is used.

2.4.1 Lincoln index

 2.12

a) What volume of oxygen does 1 dm³ 2 vol hydrogen peroxide produce?

b) What volume of oxygen does 1 dm³ 15 vol hydrogen peroxide produce?

c) What volume of oxygen does 4 dm³ 15 vol hydrogen peroxide produce?

d) What volume of oxygen does 20 cm³ 5 vol hydrogen peroxide produce?

The Lincoln index is used to estimate the number of animals in a population. It is sometimes called mark-recapture, summarising the method of collecting data. It is suitable for use with animals such as woodlice or voles but when it is practised in school, the Year 7s are often used. During a given time, such as 10 minutes, as many organisms as possible are captured. Let us call the number in the sample a and the total population A. They are marked in some way and released to mix at random. After a suitable time, such as one hour, 10 minutes are spent again collecting organisms. Let us call the number in the second sample B. Some will have been caught for a second time and they are marked. Let us say their number is b.

In summary so far:

total population = A
number in first sample = a
$\Big\}$ Proportion of population caught = $\dfrac{a}{A}$

number in second sample = B
number marked in second sample = b
$\Big\}$ Proportion of second sample marked = $\dfrac{b}{B}$

$$\frac{a}{A} = \frac{b}{B} \quad \therefore A = \frac{a \times B}{b}$$

i.e. total population = $\dfrac{\text{number in first sample} \times \text{number in second sample}}{\text{number marked in second sample}}$

2.4.2 Simpson's index

Simpson's index, D, is a measure of the biodiversity of a community, such as the invertebrates in a stream. Its measurement takes into account:

- The species richness, i.e. the number of different species.
- The species evenness, sometimes called the 'relative abundance'. If there is a similar number of organisms in each of the species present, the community has a high species evenness. But if one species is dominant, like in an oak wood, then there is a low species evenness.

There are different ways of calculating Simpson's index and your examination board may require you to use a particular formula. You are not expected to remember the formula but you are expected to be able to substitute into it, derive an answer and explain what the answer means.

Imagine a stream with four invertebrate species. The table shows the total number of organisms (N), the number in each species (n). Calculated from this are $\left(\dfrac{n}{N}\right)^2$, used in the first formula and, in the second, third and fourth formulae, $(n-1)$ and $n(n-1)$. $\Sigma n(n-1)$ shows all the values of $n(n-1)$ added together.

Species	Number (n)	$\dfrac{n}{N}$	$\left(\dfrac{n}{N}\right)^2$	$N-1$	$n(n-1)$
Mayfly nymph	12	0.31	0.10	11	132
Water shrimp	13	0.33	0.11	12	156
Stonefly nymph	11	0.29	0.08	10	110
Bloodworm	3	0.08	0.01	2	6
Total	$N = 39$		$\Sigma\left(\dfrac{n}{N}\right)^2 = 0.30$ (2 dp)		$\Sigma n(n-1) = 404$

The table below shows four different formulae, the results of the calculations and their interpretation.

Formula	Calculation	Interpretation
$D = 1 - \Sigma\left(\dfrac{n}{N}\right)^2$	$D = 1 - 0.30 = 0.70$	Higher number = higher diversity. Maximum = 1 = infinite diversity. 0 = no diversity.
$D = 1 - \dfrac{\Sigma n(n-1)}{N(N-1)}$	$D = 1 - \dfrac{404}{1482} = 1 - 0.27 = 0.73$	
$D = \dfrac{N(N-1)}{\Sigma n(n-1)}$	$D = \dfrac{1482}{404} = 3.67$	Higher number = higher diversity with no upper limit; 0 = no diversity.
$D = \dfrac{\Sigma n(n-1)}{N(N-1)}$	$D = \dfrac{404}{1482} = 0.27$	Higher number = lower diversity; Maximum = 1 = no diversity; 0 = infinite diversity;

Calculations such as these are useful for comparing biodiversity in different habitats.

For calculating the biodiversity of plants, Disney's index is commonly used, but neither this nor any of the many other biodiversity indices are required in post-16 Biology courses, so they will not be discussed here.

2.4.3 Heterozygosity index

The heterozygosity index (H) describes biodiversity within a species at the genetic level. The biodiversity of a population is calculated using the equation

$$H = \frac{\text{number of heterozygotes}}{\text{number of individuals in the population}}.$$

Homozygous plants have either red (RR) or white (WW) flowers and heterozygotes are pink (RW). Imagine a population of snapdragons, of which 300 have red flowers, 300 have white and 600 have pink. For this population,

$$H = \frac{600}{300+300+600} = 0.5.$$

2.5 The haemocytometer

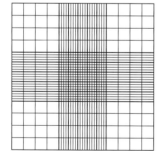

Haemocytometer slide

The haemocytometer is a modified microscope slide used to count the number of cells in a known volume of liquid. The haemocytometer slide is etched in two places as shown on the left.

The cover slip is 0.1 mm above the slide and the ruled grid covers a known area. This means that when the cell suspension is introduced into the haemocytometer chamber, you have a known volume of liquid above the grid. When the cells have settled, you count them. Often, the central area marked X on the diagram below is used for counting. Because there are two such ruled areas on the haemocytometer, the exercise can be repeated, to confirm the reliability of the counting.

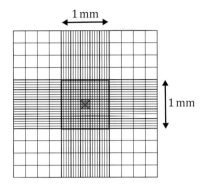

Square X is subdivided as shown below and the cells in the squares labelled A–E are counted:

Each of the squares A–E is 0.2 mm on each side ∴ each has an area $0.2 \times 0.2 = 0.04$ mm^2.

∴ the total area for the 5 squares $= 5 \times 0.04 = 0.2$ mm^2.

The depth is 0.1 mm ∴ the volume above the five squares $= 0.1 \times 0.2$ mm$^3 = 0.02$ mm^3.

If there are M cells in squares A–E, there are M cells/0.02 mm$^3 = \dfrac{M}{0.02}$ cells/mm^3.

For accuracy, the haemocytometer count should be between 70 and 100 cells. If there are too many, the solution under test would be diluted until an accurate count can be made.

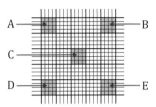

Here is a sample calculation:

A 1 in 100 dilution of cells was made and 80 were counted in the 5 squares

∴ there are $\dfrac{80}{0.02}$ cells/mm^3 in the 10^{-2} dilution

∴ there are $\dfrac{80}{0.02} \times 100 = 400\,000 = 4 \times 10^5$ /mm^3 in the undiluted suspension.

quickfire 2.13

If 70 red blood cells were counted in five squares of a haemocytometer, how many were in 1 mm^3 of the suspension?

2.6 Joints as levers

The body's joints are fixed positions around which muscles act on bones and generate movement. A joint is a pivot or fulcrum. The muscles moving a bone produce a force called the effort. The weight they move is the load. A load being moved by an effort around a fulcrum constitutes a lever so the joints of the body are levers. The relative positions of the three elements of a lever determine which class a lever belongs to:

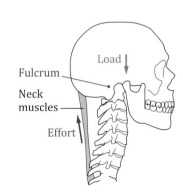

- Fulcrum in the middle = 1st order lever. The peg on the axis (the second vertebra) fits through the space in the atlas (the first vertebra). This makes a fulcrum, which supports the weight of the head, which is the load. The effort to hold up the head is provided by neck muscles.

- Load in the middle = 2nd order lever. The body does not contain second order levers but when you do push-ups, the load, which is the weight of the body, pivots around the toes, which are the fulcrum. The effort is the force produced by the arm muscles contracting when you push on the floor with your hands.

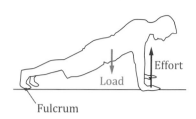

- Effort in the middle = 3rd order lever. Most of the body's levers are third order. With muscles inserted close to the fulcrum, they produce an effort which is larger than the load. This gives a considerable turning force around the fulcrum, which means the load can be moved through a large distance and can be moved quickly.

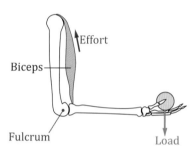

>> *Pointer*

Remember "FLE". If the fulcrum is in the middle it is a first order lever; if the load is in the middle it is a second order lever; if the effort is in the middle, it is a third order lever.

>> *Pointer*

Remember to convert mass (kg) into force (newtons).

A lever arm pivots around the fulcrum. The turning is called the torque. It depends on the force applied and how far it is from the fulcrum. A lever balances when the torque produced by the load is equal to the torque produced by the effort: $F_L \times d_L = F_E \times d_E$, where F_L and F_E are the forces of the load and effort, in newtons, and d_L and d_E are the distances of the load and the effort from the fulcrum, in metres.

Sample calculation

An adult's head has a mass of 5.0 kg and its centre of mass is 14 cm above the atlas, the first vertebra. The neck muscles can be thought of as inserting 2 cm below the atlas. If a mass of 1 kg weighs 9.8 newtons, what is the effort the neck muscles produce to hold up the head?

$$F_L = (5.0 \times 9.8) \text{ newtons}$$
$$d_L = 14 \text{ cm} = 0.14 \text{ m}$$
$$d_E = 2 \text{ cm} = 0.02 \text{ m}$$
$$F_L \times d_L = F_E \times d_E$$
$$F_E = \frac{F_L \times d_L}{d_E} = \frac{5.0 \times 9.8 \times 0.14}{0.02} = 343 \text{ newtons}$$

quickfire >> 2.14

Imagine balancing a 50 newton weight on your hand. Calculate the effort produced in your biceps if the distance from your elbow to the centre of your palm is 0.4 m and your biceps inserts 0.05 m from your elbow.

2.7 Unfamiliar mathematical expressions

Examination questions often start with several lines of information to describe what the question is about. This is called the stem of the question and it is very important that you read it all, rather than launching straight in and writing on the dotted lines, however tempting that might be. This is because the stem tells you what you need to know and it will focus your thinking. Also, it will not give you irrelevant information, so everything it says needs to be considered. Sometimes the stem will give an equation or expression you have never met before. Your teachers, text books and study guides have not been negligent. The idea is to test your approach to something new. All the information you need to answer the question is there, but you are being asked for some original thought.

Here is an example that requires you to interpret experimental results using the unfamiliar concept of assimilation number.

>> **Pointer**

All the information for calculating with an unfamiliar expression will be given to you. Don't panic.

The rate of photosynthesis may be described by the assimilation number, which is the mass in grams of carbon fixed per gram of chlorophyll a per hour. The table below gives the results of an experiment to compare the rates of photosynthesis of two geranium varieties. Interpret and explain the significance of these results.

Variety	Description of leaves	Mass chlorophyll in leaves / mg g^{-1}	Assimilation number / g carbon / g chlorophyll a / h
1	Pale green	16.0	0.007
2	Variegated	1.4	0.074

There are several points to make here that would get credit in an answer:

- The pale green leaves have more chlorophyll per gram than the variegated leaves.
- The pale green leaves have $\frac{16.0}{1.4}$ = 11.4 times more chlorophyll per gram than the variegated leaves.
- The variegated leaves have a higher assimilation number, i.e. fix more carbon per gram than the pale green leaves.
- The variegated leaves have an assimilation number which is $\frac{73.9}{1.4}$ = 52.8 times higher than that of the pale green leaves.
- Although the variegated leaves do not have chlorophyll in parts, the chlorophyll they have is more active in doing photosynthesis that the chlorophyll of pale green leaves.
- This increased activity compensates for a smaller amount of chlorophyll and synthesises enough glucose to supply their respiratory needs.

Pelargonium

>> Pointer
When comparing values, multiply or divide them to say how many times more one of them is than the other. Do not add or subtract because the answer will not describe the context of the effect.

You are unlikely to have read any of this or thought about it before, but you are expected to make inferences based on your knowledge and on your interpretation of the data.

Here is an example that requires a novel calculation.

In one year, a cow takes in 3000 kJ from each m^2 of grass eaten. From this she loses 1880 kJ in waste products and 1000 kJ in respiration. Calculate

a) the energy available for new biomass

b) the energy efficiency, given that

$$\text{energy efficiency} = \frac{\text{energy entering biomass}}{\text{energy available}} \times 100$$

a	Write the equation in words	energy available for new biomass = energy taken in − energy lost in waste 　　　　　　　　　　　　　− energy lost in respiration
	Substitute in the numbers	energy available for new biomass 　　　　　　= (3000 − 1880 − 1000) kJ
	Calculate the answer	energy available for new biomass = 120 kJ
b	Write the equation in words	$\text{energy efficiency} = \dfrac{\text{energy entering biomass}}{\text{energy available}} \times 100$
	Substitute in the numbers	$\text{energy efficiency} = \dfrac{120}{3000} \times 100$
	Calculate the answer	energy efficiency = 4%

A cow

Test yourself   2

❶ Complete the table

Substance	Passing through proximal coiled tubule	Passing through collecting duct	Quantity reabsorbed	% reabsorbed
Water	206 dm³	1.7 dm³	204.3 dm³	
Urea	206 g		206 g	
Glucose	61 g	29 g		

❷ When human blood is 100% saturated, it carries about 20 cm³ of oxygen per 100 cm³ blood. If fully saturated blood flows into muscle and loses 70% of its oxygen, what volume of oxygen is given up per 100 cm³ blood? What volume of oxygen remains in the blood?

❸ The Persian jird, *Meriones,* is a rodent and when fully grown is about 20 cm long. An individual of mass 58 g lost 6% of its body mass in 2 days on a sub-optimal diet. Calculate its mass loss and its final mass.

❹ Consider two cubes of beetroot tissue of sides of length 2 mm and 5 mm. They are weighed and incubated in distilled water for 12 hours, after which they are weighed again, having absorbed water by osmosis. Which will gain a greater percentage mass? Explain your answer, showing all the working for your calculations.

❺ Given 30 cm³ of a solution containing 150 mmol hydrogen peroxide per dm³, describe how you would dilute it to prepare a solution containing 75 mmol per dm³.

❻ The red blood cells from 10 cm³ blood were disrupted and the phospholipids that form their membranes were spread out to make a monolayer. The layer had an area of 18 m². Given that there are 7 500 000 red blood cells per mm³ of blood, estimate the area of a single red blood cell.

❼ Britain has 18 native bat species, compared with 1240 worldwide. Indonesia has 175 native bat species. Calculate the percentage bats native to Britain and to Indonesia and suggest a reason for the difference.

❽ In testing the bacterial count of various samples of ice-cream, serial dilutions were made to produce samples 10^{-1}, 10^{-2}, 10^{-3}, 10^{-4}, 10^{-5} and 10^{-6} of the original concentration. The bacteria were plated in 1 cm³ samples and cultured for 24 hours at 35 °C. The dilution 10^{-4} produced 72 colonies. What was the bacterial concentration in the original ice-cream sample?

❾ If a bacterial suspension were diluted 1000 times and gave a count of 100 in five squares of a haemocytometer, how many were in the original suspension?

Chapter 3

Graphs

 3.1

In these experiments, which is the independent variable and which is the dependent variable?

a) Volume of carbon dioxide exhaled by locusts at different temperatures.

b) Effect of different alcohol concentrations on heart rate of *Daphnia*.

c) Measuring rate of flow of water in a river and testing its correlation with the number of water shrimp.

>> **Pointer**
Check that you can spell dependent and independent.

The word 'graph' describes a way of representing data that shows the relationship between variables. The type of graph you use depends on the type of data and what you want to show. In an examination, you may be asked to draw a graph, to interpret a graph or to make measurements or calculations from it. To do any of these things, you need be clear what the different types of graph are and what they mean.

3.1 Axes

Whatever type of data you have, the graph will usually require two axes, which are lines drawn at right angles. The *x* axis is sometimes called the horizontal axis and it is used to represent the independent variable, i.e. the factor that the experiment is testing. If the factor has discrete values and only has whole numbers, such as the number of teeth in the jaws of mammals, or if it is continuous, such as the body length of salmon, the *x* axis will have a numerical scale. But if the data are categorical, e.g. lions, tigers, panthers, it will not.

The *y* axis is sometimes called the vertical axis, although as the graph is in two dimensions on a flat page, 'vertical' is not a very good word. It is, however, what everyone uses. The *y* axis carries the dependent variable. This is the factor that is measured or calculated from the readings and it depends in the independent variable. It represents the measurements or counts taken, or numbers derived from them. Numerical data is plotted on the *y* axis.

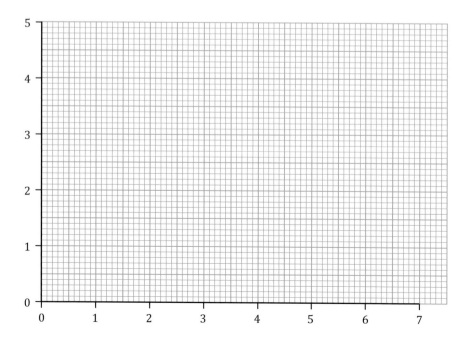

3.2 Axis scales

The scales you use on the two axes depend on the range of data that you have. Here are examples that you are likely to meet.

The commonest scales on a graph are linear. This means that on the page, equal distances represent the same value of the variable, as shown on the left. You would use linear scales if, for example, you were testing the effect of solute potential of bathing solution on mass gain by a sample of leaf tissue or the effect of temperature on the rate of reaction of amylase.

Sometimes, the range of data is so large that if you used a linear scale, the graph would have to be huge to show all the points clearly. In those cases, instead of a linear scale, a logarithmic or log scale is used. The distance on the page between two numbers on the axis is ten times the value of the preceding two numbers, as shown on the y axis on the graph below. The range 1–10 takes up the same distance on the page as 10–100, 100–1000 and so on.

In this example, the numbers are given in full, but you might see them written as powers of 10, e.g. 10^0, 10^1, 10^2, etc.

You might use a scale like this if you were testing the effect of temperature on the number of bacteria in a given volume of medium.

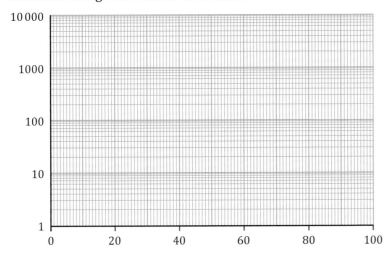

Another type of scale you might use has a log scale on the x axis, for example when you are testing a large range of concentrations. In this case, if the concentrations are all powers of ten, the graph could be plotted on linear paper, as shown below for the concentration of copper sulphate in an experiment testing the effect of its concentration as an enzyme inhibitor.

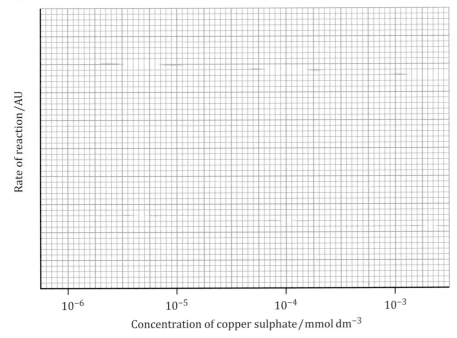

If the concentrations had intermediate values, e.g. 2×10^{-4}, 5×10^{-3} mmol dm^{-3}, however, you would need to have log paper where the log scale was on the x axis, as shown on page 54. It is not commonly used.

>> *Pointer*

You can use a linear scale or a log scale, but it must be uniform all the way along the axis.

quickfire >> 3.2

Look at these data tables and suggest suitable axes and scales for the graphs you could draw:

a)

Dominant hand	Number of people / 10 000
Left	1100
Right	8880
None	20

b)

Temperature / °C	Mass starch digested by amylase / g
10	0.8
30	9
50	14
70	19
90	11

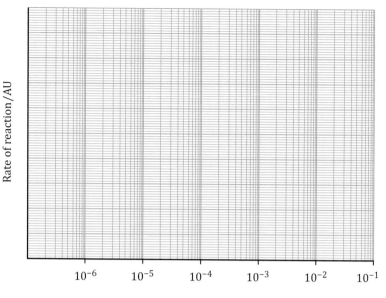

Concentration of copper sulphate / mmol dm^{-3}

Log scales do not have a zero. If you carried on going further to the left on the set of axes above, you would see 10^{-7}, 10^{-8}, 10^{-9}, etc., getting smaller all the time, but you would never reach zero.

3.3 Types of data

Data can come in different forms and each form is associated with a different type of graph.

3.3.1 Categorical data

Categorical data has data in different categories. A graph might show how much glucose is found in various fruits. You cannot draw a scale to demonstrate the different fruits, e.g. apple, orange, pear, so each fruit is a different category. In this case, you could display the information in a bar graph. The independent variable is the type of fruit, so that goes on the *x* axis. The glucose concentration is the dependent variable and so that is plotted on the *y* axis. The *y* axis has a standard scale, which you choose depending on the results.

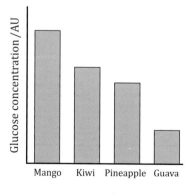

Type of fruit

>> *Pointer*

With categorical data, the bars on the bar chart do not touch.

The bars are not drawn touching each other, to illustrate the fact that they are separate categories. There are equal spaces between them.

3.3.2 Discrete data

Discrete or discontinuous data are data that occurs in whole numbers, such as the number of puppies in a litter or petals on buttercup flowers. There are numbers, so a scale is possible, but there are no fractions of the values. The data can be plotted as a frequency chart as shown on the right.

In biological systems, characteristics which generate discrete or categorical data are those controlled by single or a small number of genes. Examples include the ABO and Rhesus blood group systems in humans and petal colour in *Antirrhinum* flowers.

Populations with such characteristics are described as showing discontinuous variation.

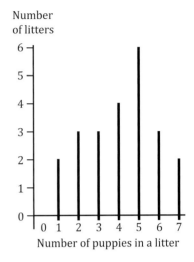

3.3.3 Continuous data

Continuous data can take any value within a certain range, e.g. human height. There are no set values that human height has and if you measure a population, you will find a complete gradation within the range. When this type of data is plotted, it can be done as a frequency histogram. You define classes and count the frequency of values within each class. That frequency is what is plotted on the y axis. As each class is continuous with the next, the bars are joined. Here is an example:

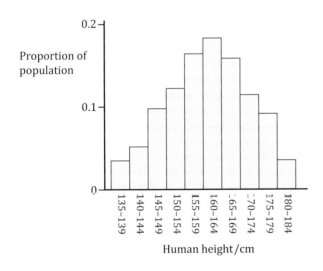

Continuous data with a class size of 5 cm

In this case heights are to the nearest cm and the classes were chosen to be 5 cm, e.g. 150–154 cm, but you can choose the class, depending on the range and the raw values of the data. You can choose how many bars you have by deciding on the class size. If the class sizes for the data plotted above were 10 cm instead of 5 cm, the same data above would look as in the diagram at the top of page 56.

If plotted on the same vertical scale, as shown here, each bar is double the width and each bar's height is the sum of two adjacent bars on the previous histogram. In each case, though, the total adds up to 1 because it represents the whole population measured.

The vertical axis here is labelled 'proportion of population' but it could have been plotted and labelled as frequency or percentage in population.

quickpire » 3.3

Are the types of data categorical, discrete or continuous?

a) Birth weight of human babies.
b) Human blood groups.
c) Number of chromosomes in plants of the cabbage family.

» Pointer

In a frequency histogram, the classes are continuous so the bars of the histogram touch each other.

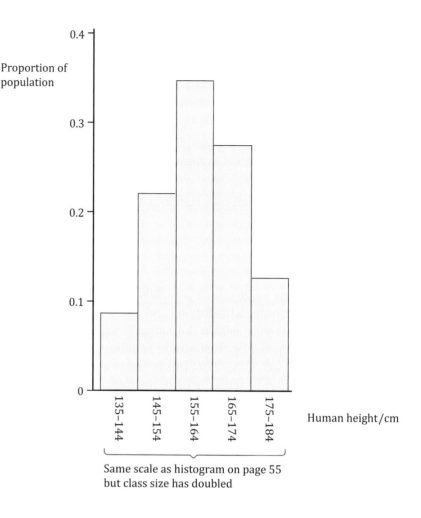

Same scale as histogram on page 55
but class size has doubled

Continuous data with a class size of 10 cm

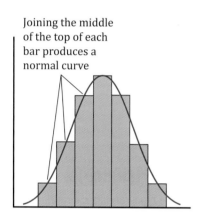

Joining the middle
of the top of each
bar produces a
normal curve

Characteristic with continuous
variation showing normal distribution

》 Pointer

Make sure you know if the
data are categorical, discrete or
continuous.

As the classes become smaller, the bars become narrower. If this is taken to
the extreme, and each bar is infinitely narrow, the tops of the bars can be
joined to make a smooth curve. In an ideal case, the curve is symmetrical with
a single peak. For the values in this curve, the mean, median and mode are all
the same. This curve is called a normal curve and it represents what is called
a normal distribution. It is sometimes called a Gaussian distribution after the
mathematician who first defined it. In biological systems, the distribution
in a population of characteristics controlled by many genes as well as the
environment is described by normal curves. Another example is human weight.
Such characteritics are said to show continuous variation, as they can take any
value within a given range.

3.4 Population pyramids

A population pyramid is a combination of two histograms placed sideways
and joined at their bases, showing males and females in certain age brackets.
The example on page 57 shows the per cent of males and females in 5-year age
brackets in a population. Along the horizontal axis, the scale may be in millions
of people or, as shown here, the percentage of the population. The shape of the
pyramid shows the current population structure and gives information about the
past and allows tentative predictions for the future.

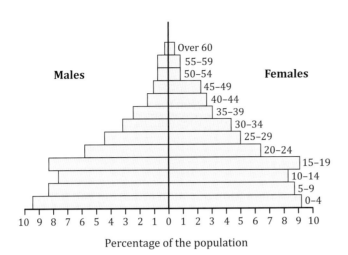

Percentage of the population

≫ Pointer
Think of a population pyramid as two histograms joined at their bases.

You might be asked to read from the pyramid, for example, the percentage or the number of females in a certain age group, e.g. in the pyramid shown above, 9% of females are represented by the 15–19 year age group. But if you are asked for both males and females, the sum from both sides of the pyramid must be given, e.g. percentage of people in the 15–19 year age group = 9 + 8.5 = 17.5%.

quickfire ≫ 3.4

a) What is the percentage of the males in the population shown by the population pyramid that is 0–4 years old?

b) What percentage of the population is between 30 and 34 years old?

3.5 Line graphs

Line graphs are used to display experimental data when both the independent and dependent variable can be plotted on a numerical scale. When given data to plot on a line graph, choices must be made about presentation and it is logical to consider the axes and then the scales. How the axes and scales are labelled is crucial in ensuring that the graph accurately represents the data. Here are the steps to go through.

3.5.1 Axis choice

The independent variable is plotted on the x axis. If you are not sure which is the independent variable, ask yourself what the experimenter has chosen to vary and has chosen the values for. In a table of data, the numbers for this variable are likely to be regularly spaced, e.g. distance along a transect given as 20, 40, 60, 80, 100 m, or in a sequence, e.g. nitrate concentration in a culture medium given as 0.0, 0.05, 0.10, 0.15, 0.20 mol dm^{-3}.

The dependent variable is plotted on the y axis. This is what has been counted or measured or has been calculated from what has been counted or measured. If you are not sure which is the dependent variable, ask yourself which set of numbers is not regularly spaced. That is likely to be the dependent variable.

When you draw the axes, they do not need little arrows on the end. The arrows mean that the scale carries on in the same way to infinity. That is acceptable in mathematics because numbers behave in a predictable way. But living organisms do not, so the arrow is not appropriate.

≫ Pointer
The values you choose go on the x axis. The values you measure, count or calculate go on the y axis.

quickfire ≫ 3.5

For the experiments described, state which data set you would plot on the x axis and which you would plot on the y axis:

a) Testing the effect of pH on the rate of digestion of casein in milk by trypsin.

b) Testing the effect of time at room temperature on the bacterial count of milk.

c) Testing the effect of light intensity on the area of ground ivy leaves.

» Pointer

Don't forget the units when labelling the axes.

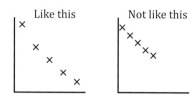

Like this   Not like this

3.5.2 Axis labels

There is no need to write x and y on the axes. They should only be labelled with the name of the variable and its units. Give complete information in the axis title, such as 'time for methylene blue to decolorise' rather than just 'time'. The current convention is to write an oblique between the name of the variable and its units, e.g. temperature /°C, rather than putting the units in brackets, as was done in the past.

3.5.3 Scale choice

a) **Graph size**. In an examination question requiring you to plot a graph, great care is taken in choosing the size of the graph paper provided. Your task is to choose a sensible scale that uses as much of the graph paper as you can for plotting the points. If there is a range of values over many orders of magnitude, such as in data showing the increase in the number of bacteria in a population, a logarithmic scale will be more suitable than a linear scale.

b) **The scale does not have to begin at zero.** Taking each variable in turn, look at the range of data and make sure the lowest extent of the scale is just below the lowest reading and the highest point of the scale is just above the highest reading. By counting squares, you can see how to fit that range in, taking as much of the paper as you can. But do take care to use sensible numbers. For example, if the data range is 42–98, you might choose to have your axis scale in the range 40–100. You would mark on the axis in 10 s, i.e. 40, 50, 60, 70 80, 90, 100. Make sure you can fit that on to the paper. If there is a great deal of space, you may choose to mark in 5s instead. If you do not have enough space, you may choose to mark in 20s or, perhaps, turn the graph paper through 90° and see if it fits that way.

c) **The scale must be uniform** along its whole length. You may have several results between 0 and 1 and not so many between 1 and 10 but the scale 0–10 must be uniform. Sometimes, students have an expanded scale between 0 and 1 to show those points more clearly, but then the graph has lost its meaning as the separation of the points is not uniform.

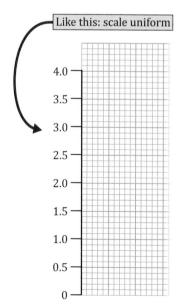

Like this: scale uniform

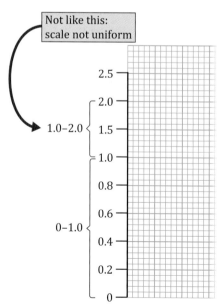

Not like this: scale not uniform

d) There must always be a number at the origin, although it does not have to be 0. However, if both axes begin at 0, you could write a 0 on each axis or write a big O at the origin. This must be very clear though

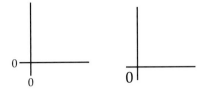

e) **Scale breaks.** Sometimes you see zero at the start of a scale followed by a scale break. This is acceptable but it must be in the correct place, i.e. between the origin and the first scale number. If the break comes between two numbers on the scale, the scale is no longer uniform.

3.5.4 Points

It is conventional to make the point as an X which is clearly written and easy to see. A + sign for a data point may be obstructed by the grid line of the paper and a point may be made invisible when the line is drawn.

3.5.5 Line

a) **Point-to-point.** In most cases, in Biology, a ruler is used to join point-to-point. This is because only the data points are real. If you draw a curve of best fit, what happens between the points is just a construct of your imagination. The point-to-point joining says that it is not known what happens between the points, but it is possible that the results would follow the general direction of the line. When drawing the line, you should use a sharp pencil and make sure the line goes precisely through the centre of the X that is the data point.

b) **Line of best fit.** If the graph is being used to read an intercept, such as in an experiment to determine the water potential of plant tissue, a line of best fit is suitable. In this case, the reason is that if the point-to-point method were used, then to read the intercept, only two points would be needed, the two either side of the intercept. But there is intrinsic uncertainty in any data point, even if it is calculated as a mean of several readings. So the reliability of the intercept is greater if a line of best fit is used, as it takes into account all of the points, not just two.

It is possible to calculate the co-ordinates the line of best fit goes through. The sum of the squares of the distances of each point from the line on one side of it must equal the sum of the squares of the distances of the points on the other side. When the computer plots a regression line, these calculations determine where the line is placed. But this is about Biology, not Mathematics, and the calculations are repetitive. In a Biology examination, it will be sufficient for you to judge by eye where the line, or indeed curve, of best fit should fall. The points either side of the line should seem equally weighted so that they all have the equivalent influence on the position of the line, as shown here:

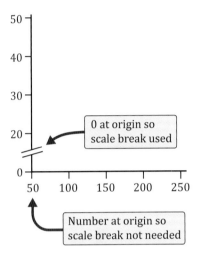

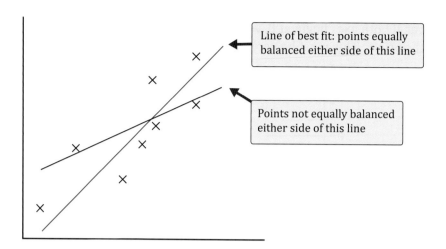

3.5.6 Key

If you use the same set of axes for more than one set of data and therefore have more than one line on the graph, make sure you label each line, or draw them differently, e.g. in different colours, with different shaped data points or with one line dotted. Give a key so that it is absolutely clear what each line represents.

3.6 Interpreting graphs

3.6.1 Describing graphs

You may be asked to describe a graph. There are specific aspects you should consider in your answer. These will be illustrated by describing the graph below, which shows the effect of glucose concentration on the rate of a reaction:

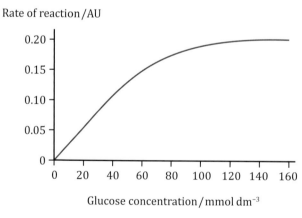

a) First give a general description – overall, as the concentration of glucose increases, the rate of reaction increases.

b) Give details:

   i. Without glucose, there is no reaction, i.e. the line goes through the origin.

   ii. Between 0 and 60 mmol dm^{-3} glucose, the rate of reaction produced is directly proportional to the concentration of glucose.

   iii. Above 140 mmol dm^{-3} the rate of reaction remains constant.

c) Process numerical data, e.g. at 60 mmol dm^{-3} glucose, the rate of reaction is approximately double that produced at 30 mmol dm^{-3} glucose. This does not have to be a precise calculation and for this purpose, a rough estimate from examining the graph is suitable. You could use symbols in your answer and write 'at 60 mmol dm^{-3} glucose the volume produced is ≈ ×2 of that produced at 30 mmol dm^{-3}'.

Points to remember

1 **Describe the dependent variable:** the description of a graph should be in terms of the behaviour of the dependent variable, in this case the rate of reaction. You would write 'at concentrations above 140 mmol dm^{-3} glucose, the rate of reaction remains constant'. This is preferable to 'the graph plateaus' or 'the line is horizontal', because neither of those phrases describes the dependent variable.

2 **Don't describe individual data points:** the graph below shows the decline in a population of bloodworms along a stream, away from a sewage outlet. This decrease might happen if the concentration of oxygen increases. At 80 m the number decreases but at 90 m it increases. The data show that this alternating pattern continues up to 110 m.

When you describe a graph, you do not need to say what happens for each data point. That is, after all, in the results table and shown on the graph. But you should summarise by writing that the number of bloodworms decreases up to 80 m from the sewage outlet and between 80 m and 110 m the number fluctuates.

If the decrease is linear, you could describe the relationship as inversely proportional. The symbol for this is $\frac{1}{\propto}$. Similarly, a linear increase might be described as proportional, with the symbol ∝. If the line goes through the origin, it may be described as directly proportional.

> **Pointer**
>
> When describing a graph, talk in terms of the dependent variable, not 'the line' or 'the graph'.

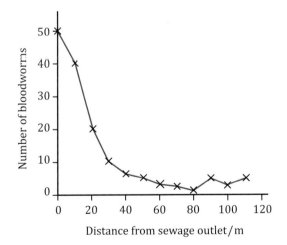

3 **Remember the units on the *y* axis:** this graph shows the number of sheep on an island where the population has reached its carrying capacity. If asked to read how many sheep can be supported on this island, check the *y* axis label.

Do not forget to multiply by 1,000,000.

You cannot have 1.5 sheep.

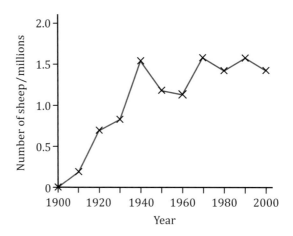

4 **Different scales:** Sometimes two plots have different scales on the *y* axis. The graph below shows the populations of yeast and *Paramecium*, its predator. If asked to describe the populations, note that the scales labelled for the two species are different. It is therefore not correct to say that at the peaks of population, there are more *Paramecium* than yeast.

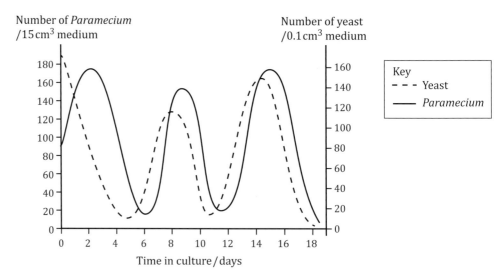

5 **Note if the scale is linear or logarithmic:** If data points are in a straight line and the vertical axis has a linear scale, the increase is linear. But if, as on a curve showing the numbers in a population of bacteria over time, the vertical axis has a log scale, the increase is exponential. This is described further on page 71.

3.6.2 Numerical analysis of graphs

a) Reading an intercept

You may be asked to give a value of one variable when the value of the other is zero. This means you have to read the value where the line of the graph crosses the axis. This point is the intercept, either the x-intercept or the y-intercept, depending on which axis you are reading from.

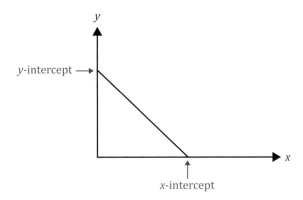

A straight line is defined by the equation $y = mx + c$, where m = gradient and $c = y$-intercept. This equation can be applied to a situation in biology where a relationship is linear. Examples include the length of the femur in relation to the height of adult humans, where the gradient is positive, and competition between two species in a habitat, where the gradient is negative.

The equation can be applied to find the maximum rate, V_{max}, of an enzyme-controlled reaction, in a graph called a Lineweaver–Burk plot. The relationship between substrate concentration (S) and rate of reaction (V) is shown in the graph, below left. If, instead of plotting V against S, $\frac{1}{V}$ is plotted against $\frac{1}{S}$, a straight line is seen, below right. The line is described by the equation $y = mx + c$.

In this case, c, the y-intercept, is the rate of reaction when $\frac{1}{S} = 0$. When S is theoretically infinite, the rate of reaction will be at its highest (V_{max}). This is when $\frac{1}{S} = 0$ so the value of the y-intercept is $\frac{1}{V_{max}}$. If, for example, the y-intercept is at 0.01 min AU^{-1}, then $\frac{1}{V_{max}} = 0.01$ min AU^{-1}, and $V_{max} = \frac{1}{0.01} = 100$ AU min^{-1}.

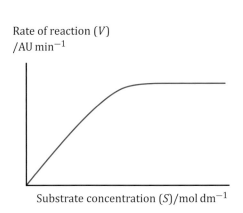

Rate of reaction (V) /AU min^{-1}

Substrate concentration (S)/mol dm^{-1}

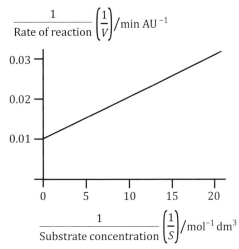

$\frac{1}{\text{Rate of reaction}} \left(\frac{1}{V}\right)$/min AU^{-1}

$\frac{1}{\text{Substrate concentration}} \left(\frac{1}{S}\right)$/mol^{-1} dm^3

>> **Pointer**

At the x-intercept, the value of the dependent variable = the value of the independent variable at the origin. At the y-intercept, the value of the independent variable = the value of the dependent variable at the origin.

The volumes of oxygen and carbon dioxide evolved and used in the respiration and photosynthesis of a molecule of glucose are equal, as the equation indicates:

$$6CO_2 + 6H_2O \xrightleftharpoons[\text{respiration}]{\text{photosynthesis}} C_6H_{12}O_6 + 6O_2$$

There is a critical light intensity at which respiration and photosynthesis are in balance and no gas exchange occurs. This light intensity represents the plant's compensation point. It is the x-intercept in a graph showing evolution or absorption of carbon dioxide against the light intensity. Shade plants, such as wood sage, have adapted to low light intensity and are able to perform photosynthesis efficiently at low light intensity. They have a lower compensation point than sun plants, such as celandines, that have adapted to a higher light intensity.

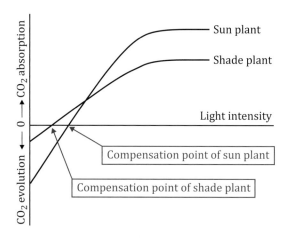

quickfire 3.6

a) What is the percentage saturation of haemoglobin at a partial pressure of oxygen at 4 kPa?

b) What is the partial pressure of oxygen when the haemoglobin is 50% saturated with oxygen?

c) What is the partial pressure of oxygen when the haemoglobin is 90% saturated with oxygen?

》 Pointer

If you have to read from a graph, draw on it and make sure the lines you draw are parallel with the gridlines.

b) **Reading off a graph:** Here is an example where you are asked to find the per cent oxygen saturation of haemoglobin at a partial pressure of oxygen which is 5 kPa. The graph is a sigmoid curve.

It is possible by eye to read along the x axis as far as 5 kPa, estimate where the graph line intercepts 5 kPa and then estimate the per cent saturation on the y axis. You may even get the right answer but if you draw on your graph, you know that your answer is correct.

Use a ruler and make sure your lines are parallel with the grid lines, as shown. It may take longer but it will give you the correct answer.

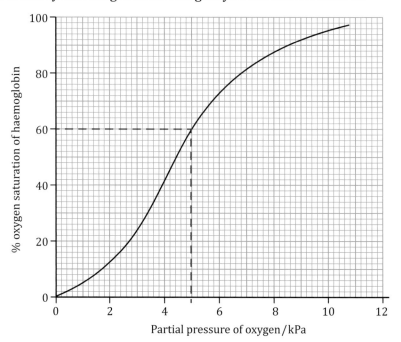

You may be asked to calculate difference in percentage saturation of haemoglobin when blood flows between parts of the body with different partial pressures of oxygen. The graph below shows two curves. Using curve ①, find the percentage change in haemoglobin saturation when blood flows from a part of the body with 8 kPa partial pressure of oxygen to an area at 3 kPa partial pressure of oxygen.

From curve ①:

% saturation at 8 kPa = 94%

% saturation at 3 kPa = 20%

∴ change in saturation = 94 − 20 = 74%

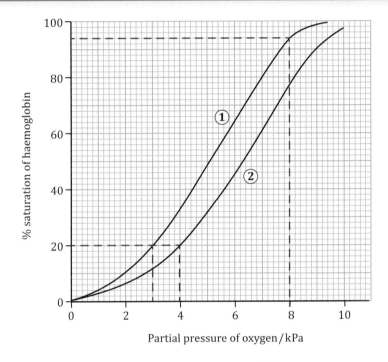

% saturation of haemoglobin

Partial pressure of oxygen / kPa

The question may ask you to convert this into an actual volume of oxygen. You can do this because you would be told that when haemoglobin is 100% saturated, it carries 110 cm³ oxygen per 100 cm³ blood. Working from first principles:

100% saturation ≡ 110 cm³ oxygen

∴ 1% saturation ≡ $\dfrac{110}{100}$ = 1.10 cm³ oxygen

∴ 74% saturation ≡ 74 × 1.10 = 81.4 cm³ oxygen
i.e. 81.4 cm³ oxygen are released.

The Bohr effect describes the observation that the per cent saturation of haemoglobin with oxygen is lower at a higher partial pressure of carbon dioxide. Curve ① shows the change in saturation at 3 kPa partial presure of carbon dioxide and curve ② shows it at 11 kPa carbon dioxide. You could be asked to calculate the volume of oxygen released when blood at 8 kPa partial pressure of oxygen and 3 kPa partial pressure of carbon dioxide, such as in the alveoli, flows into a part of the body at 4 kPa oxygen and 11 kPa carbon dioxide, such as in respiring tissues.

Blood at 3 kPa partial pressure of carbon dioxide is shown by curve ①. From the curve, haemoglobin at 8 kPa partial pressure of oxygen has an oxygen saturation of 94%.

Blood at 11 kPa partial pressure of carbon dioxide is shown by curve ②. From the curve, haemoglobin at 4 kPa partial pressure of oxygen has an oxygen saturation of 20%.

$\therefore$ change in saturation = 94 – 20 = 74%

100% saturation $\equiv$ 110 cm³ oxygen

$\therefore$ 1% saturation $\equiv$ 1.10 cm³ oxygen

$\therefore$ 74% saturation $\equiv$ 74 × 1.10 = 81.4 cm³ oxygen,
i.e. 81.4 cm³ oxygen are released.

quickfire ≫ **3.7**

What is the difference in mass of sugar produced between 5 and 10 minutes?

c) **Calculating a difference**: a more sophisticated version of reading from a graph might ask you to calculate the difference in the mass of reducing sugar produced after incubating starch with amylase for 5 minutes and 15 minutes.

Once again you have to read off the graph, but you will have to make two readings, and lines for both should be drawn, as shown below. The smaller reading on the y axis must be subtracted from the larger to give the difference:

Mass difference = mass at 15 min – mass at 5 min

= 9.0 – 3.5

= 5.5 µg ← Don't forget the units.

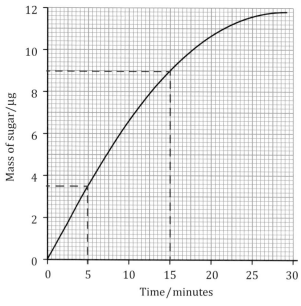

The next level of complexity asks you to find a percentage increase or decrease. The wording is crucial here so these two tasks will be considered separately. Using the graph above.

d) **Percentage increase:** if you are asked to calculate the percentage mass increase between 5 minutes and 15 minutes, first you must find the actual increase. This was done above. As it is a per cent increase you have to find the actual increase expressed as a percentage of what was there at the start of the period in question:

$$\text{\% increase} = \frac{\text{actual mass increase}}{\text{initial mass}} \times 100\%$$

$$= \frac{\text{mass at 15 minutes} - \text{mass at 5 minutes increase}}{\text{mass at 5 minutes}} \times 100\%$$

$$= \frac{9.0 - 3.5}{3.5} \times 100\% = 157.1\%$$

Having an increase of more than 100% may seem counter-intuitive, but if you look carefully at the graph, you can see that between 5 minutes and 15 minutes, the mass of sugar produced more than doubles. Double the mass would be a 100% increase, and this is more, so the answer makes sense.

e) **Percentage decrease:** if you are asked to find a percentage decrease for a value that has gone down, the same principle applies, but this time the difference will be divided by the initial, higher value. The graph below shows the length of nettle leaves at different light intensities. The task is to calculate the percentage decrease in length between 500 lux and 4000 lux.

$$\text{\% decrease} = \frac{\text{actual decrease}}{\text{mean length at 500 lux}} \times 100\%$$

$$= \frac{\text{mean length at 500 lux} - \text{mean length at 4000 lux}}{\text{mean length at 500 lux}} \times 100\%$$

$$= \frac{52.5 - 10}{52.5} \times 100\% = \frac{42.5}{52.5} \times 100\% = 81.0\% \ (1dp)$$

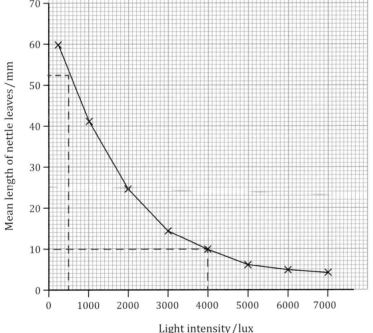

Nettle leaves

This result says that at 4000 lux, which is high light intensity, the nettle leaves are 80% smaller than at 500 lux, a low light intensity. As this is in the context of Biology, you may be asked to explain why. Most answers would explain that a leaf needs more area over which to gather light when the intensity is low. A higher quality answer would go on to relate this to the evolutionary advantage of growing to the right size. If the leaf had alleles that made it grow too big in high light, it would be wasting energy that would otherwise be used to reproduce, and such plants would be selected against. Similarly, a leaf that had alleles that did not let it grow enough at low light intensity would not capture enough light to generate enough sugar to release enough energy for its life processes and it would also be selected against.

quickfire 3.8

Calculate the percentage increase in sugar produced between 10 and 15 minutes.

>> *Pointer*

When you calculate a percentage increase or decrease, find the actual increase or decrease first. Then divide by what you started with and multiply by 100 to get the percentage.

quickfire 3.9

Find the percentage decrease in nettle leaf length between 3000 lux and 500 lux.

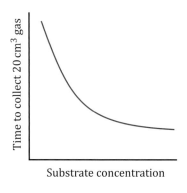

Time to collect 20 cm³ gas / Substrate concentration

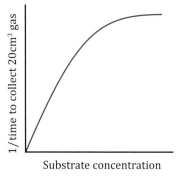

1/time to collect 20cm³ gas / Substrate concentration

» Pointer

Gradient of a graph =
change in y
change in x

3.6.3 Rates of reaction

Rate and time

A rate is a measurement of change in a given time. If there is a greater change in a given time then the rate is faster and if there is less change in a given time, the rate is slower. So rate is inversely proportional to time, i.e. rate $\propto \frac{1}{time}$. This is a useful way of thinking about a rate of reaction when, for example, you are collecting the product of a reaction over a period of time. If you were to measure the time taken to collect a given volume of gas produced at, for example, different substrate concentrations, the time would decrease and then remain constant as the concentration was no longer the limiting factor as shown on the left.

If instead of time, you plot 1/time, which is proportional to rate, you see the standard graph described by enzyme theory.

Rate and gradient of a straight line

In typical experiments, the rate of a reaction could be measured by finding the mass of lipid digested by lipase in a second, or the volume of oxygen generated in a minute by catalase acting on hydrogen peroxide. If you plotted results for either of these examples on a graph, time would be plotted on the x axis, as it is the independent variable.

The way to calculate a gradient from a graph is $\frac{\text{change in } y}{\text{change in } x}$.

Sometimes the symbol Δ is used to mean 'change of' so gradient $= \dfrac{\Delta y}{\Delta x}$

The unit for the gradient is therefore units for x / units for y and because x is time in this example, the gradient is a measure of the rate.

The graph on page 66 shows sugar production over 30 minutes. Here it is again.

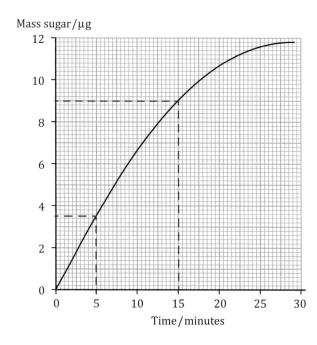

Between 0 and 15 minutes, the line is approximately straight, which means sugar is being produced at an approximately constant rate. The rate of sugar production is the gradient of the line. To find the rate of sugar production over the first 15 minutes, the gradient of the line must be calculated:

$$\text{Gradient} = \frac{\text{change in } y}{\text{change in } x}$$

$$= \frac{\text{mass at 15 minutes} - \text{mass at 0 minutes}}{15 - 0}$$

$$= \frac{9.0 - 0}{15} = \frac{9.0}{15} = 0.6 \text{ (1dp)}$$

∴ rate of sugar production over the first 15 minutes = 0.6 µg min^{-1}

You may be asked to find a rate between two different times. To find the rate between 5 minutes and 15 minutes the gradient must be found as above. Once again you have to read off the graph, but you will have to make two readings and the lines for both should be drawn, as shown. The smaller reading on the y axis must be subtracted from the larger to give the difference:

$$\text{Gradient} = \frac{\text{change in } y}{\text{change in } x}$$

$$= \frac{\text{mass at 15 minutes} - \text{mass at 5 minutes}}{15 - 5}$$

$$= \frac{9.0 - 3.5}{10} = \frac{5.5}{10} = 0.6 \text{ (1dp)}$$

∴ rate of sugar production between 5 and 15 minutes = 0.6 µg min^{-1}

As the line on the graph is approximately straight, the two calculations above give the same answer.

Rate and gradient at a point

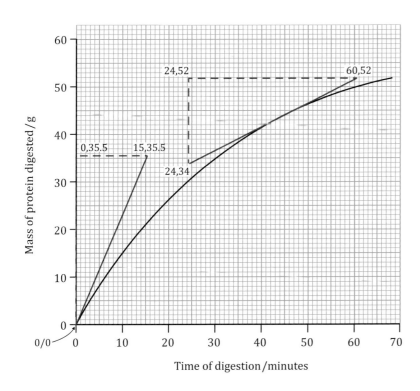

quickfire ≫ **3.10**

Using the graph on the left, find the average mass of protein digested between 10 and 20 minutes.

If the line were not straight, to find a rate at any one point in time, the gradient of a tangent at that time would be calculated. The graph above shows the results of an experiment testing the mass of protein digested over 68 minutes.

Tangents have been drawn at 0 min and at 40 min and readings made to calculate the gradient at those times:

At 0 minutes: rate of digestion $= \dfrac{\Delta y}{\Delta x} = \dfrac{35.5 - 0}{15 - 0} = \dfrac{35.5}{15} = 2.4 \, \text{g min}^{-1}$

At 40 minutes, rate of digestion $= \dfrac{\Delta y}{\Delta x} = \dfrac{52 - 34}{60 - 24} = \dfrac{18}{36} = 0.5 \, \text{g min}^{-1}$

You might be asked to make such calculations and be asked to comment on the significance of the results. You could explain that this calculation illustrates the observation that reactions are fastest at the start of a reaction and they subsequently slow down. As the substrate is used up, enzyme–substrate complexes form less frequently and so there are fewer reactions per unit time.

3.6.4 Population growth curves

A population growth curve shows how many individuals there are in a population over a period of time.

The growth curve below shows a population of deer. In real populations there is a fluctuation in numbers and the carrying capacity is the number around which the population fluctuates. In this example: carrying capacity = 1900 deer.

The growth rate of the population between 0 years and 50 years can be calculated by reading the population at those times and dividing by 50:

Rate of population growth $= \dfrac{\text{population at 50 years} - \text{population at 0 years}}{50}$

$= \dfrac{2150 - 50}{50} = \dfrac{2100}{50} = 42 \text{ deer/year.}$

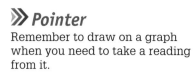

Pointer

Remember to draw on a graph when you need to take a reading from it.

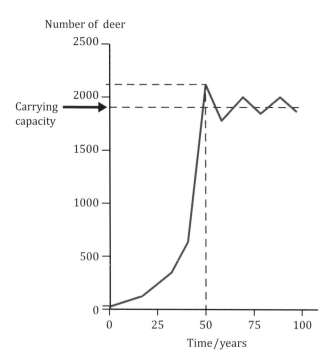

A one-step growth curve is shown below. It is called that because the graph shows a step up in numbers from the initial population to the carrying capacity. You may be asked to read the carrying capacity from the growth curve or to calculate the rate of increase of the population. The diagram shows a stylised growth curve with phases marked:

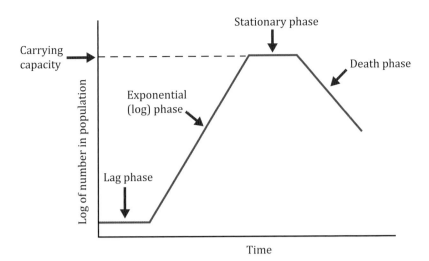

》 Pointer

A graph which is a straight line represents an exponential increase if it is plotted on a log scale.

This shape of growth curve describes the change in numbers of a population of many different organisms, including bacteria and elephants. Other types of growth curve are seen and the actual shape depends on the biology of the organism and on its interaction with its environment. The y axis here is labelled 'log of number in population'. In many populations studied in Biology, such as bacteria, the range of numbers is very large and may extend over several orders of magnitude, e.g. from 10 to 1 000 000. This huge range would make a linear scale hard to plot and hard to read. So logs are used. The phase on the graph labelled 'Exponential (log) phase' is the most rapid phase of population growth when there is no environmental resistance. A population increase is exponential if it doubles in number per unit time. If the graph has a linear scale, the line would be a curve of increasing steepness; but with a log scale, the line is straight, as on this graph.

The graph below shows a population of bacteria over 15 hours.

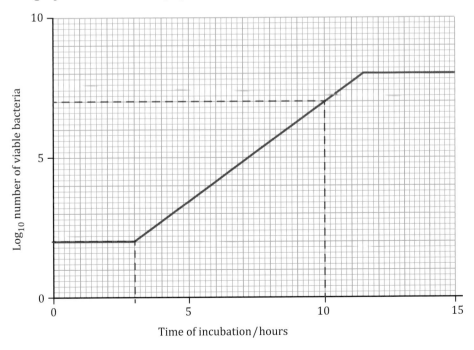

To calculate the increase in the number of bacteria between 3 and 10 hours, you would subtract the number at 3 h from the number at 10 h. But if the vertical axis has a log scale, you have to find the antilog first:

Increase in population = number at 10 h − number at 3 h

From the graph, increase in population = antilog 7 − antilog 2

= 10 000 000 − 100

= 9 999 900

Microbiologists use bacterial growth curves to find the 'experimental growth rate constant', which is the number of generations, that is, the number of divisions, one bacterium would go through, per unit time. You would not be expected to remember the equation for calculating this, but you might be required to read a graph and plug numbers into the equation. The experimental growth rate constant is given the symbol k and is calculated from the equation

$$k = \frac{\log_{10} N_2 - \log_{10} N_1}{t \times \log_{10} 2}$$

where N_2 = the final number of bacteria, N_1 = the initial number of bacteria, t = time and $\log_{10} 2 = 0.301$.

You do not need to understand the mathematics behind this equation and you do not need to know how to use logarithms. All you have to do is read from the graph and substitute into the equation.

The graph on page 71 shows the number of viable cells in a bacterial culture over 15 hours. To find the experimental growth rate constant between 3 and 10 hours, you read N_1 and N_2 from the graph and substitute:

$N_2 = \log_{10}$ number at 10 h $= 7$

$N_1 = \log_{10}$ number at 3 h $= 2$

$t = (10 - 3) = 7$ h

$$k = \frac{\log_{10} N_2 - \log_{10} N_1}{t \times \log_{10} 2}$$

$$= \frac{7 - 2}{7 \times 0.301}$$

$$= \frac{5}{2.107} = 2.37 \text{ generation/h}$$

If you were asked the number of whole generations, you would round down to 2.

3.6.5 Spirometer traces

A spirometer produces a trace showing the volumes of air inhaled and exhaled during breathing. The volumes are recorded on a rotating drum called a kymograph. Positive values show air exhaled and negative values show air inhaled. The diagram on the left is an idealised version of the trace, but notice that, unlike normal, the y axis has positive values below the x axis and negative values above. Four values are shown:

① The tidal volume is the volume of air exchanged in normal breathing.

② The vital capacity is the maximum volume of air that can be inhaled or exhaled.

③ The residual volume is the volume remaining even after exhaling as much as possible.

④ The total lung capacity is the total volume the lungs can hold. It is the sum of the vital capacity + residual volume.

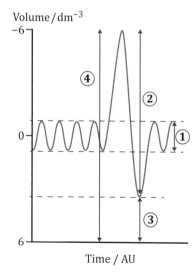

Volume / dm⁻³

Time / AU

A real spirometer trace looks more like the diagram on the right, which shows maximum inspiration and maximum expiration, with tidal breathing in between. The vertical axis shows volume, so by reading the scale, it is possible to read from the vertical axis the maximum volumes inspired and expired, giving the vital capacity, ② on the drawings, and to measure the tidal volume, ①.

The second diagram on the right is part of another trace. The horizontal axis shows time and the vertical axis shows volume, although the scale is not given here:

The diagrams can be used to make several readings and calculations:

a) How long a breath takes (10 s).

b) How many breaths are made in a minute (6 in the 60 s indicated). This is the breathing rate.

c) The tidal volume is shown.

d) The breathing rate can be used to find the ventilation rate.

  Ventilation rate = breathing rate × tidal volume

e) Oxygen consumption is the volume of oxygen absorbed into the blood in one minute. It is shown in Diagram 2. The negative gradient shows that the oxygen available in the spirometer is being used up.

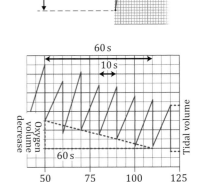

3.6.6 ECG traces

The trace below shows a normal ECG, annotated to show its characteristics.

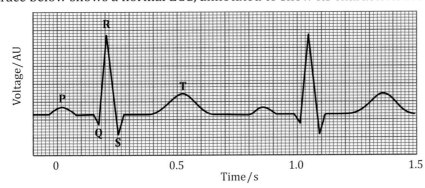

Length of cycle = time between equivalent points on two traces, e.g. from one R to the next.

Cycle length = 1.05 − 0.21 = 0.84 s.

From first principles:

In 0.84 s is one cycle

∴ in 1 s are $\dfrac{1}{0.84}$ cycles

∴ in 60 s are $60 \times \dfrac{1}{0.84} = 71.4$ cycles

So the heart rate = 71.4 bpm.

The P–R interval is the delay between the atria and the ventricles contracting. In this example, P–R interval = 0.21 − 0.02 = 0.19 s.

You may be shown abnormal traces for comparison. You might be expected to recognise a cycle time that was shorter or longer than the normal and to describe abnormalities in the QRS complex, as shown:

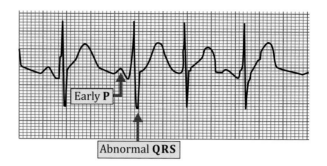

Early P

Abnormal QRS

3.6.7 Kite diagrams

An environmental gradient can be investigated by making a belt transect and assessing the change in plant populations along it. A gridded or point quadrat can be used to estimate per cent frequency of species along the transect, as discussed on page 34. The gridded quadrat can also be used to find the per cent area cover, by estimating how many complete squares of the grid are covered by the species in question. If the grid is 10 × 10, giving 100 squares in all, each square represents 1% of the total so the total number of squares covered = percentage area cover. Some grids are 5 × 5, giving 25 squares so each represents 4% of the area, making the estimation less accurate than that of the 10 × 10 grid.

The smaller the squares, the more accurate the estimation. The per cent area cover of the species may change along the transect and this change can be shown in a kite diagram. The table on the left shows the percentage area cover of ground ivy along a transect from underneath an oak tree with a 12 m diameter canopy, into an open field.

Distance along transect / m	% area cover ground ivy
0	80
2	80
4	72
6	56
8	32
10	20
12	0
14	8

In a kite diagram, the percentage area cover is plotted symmetrically above and below a horizontal axis representing the transect. Half the value will be above the axis and half will be below. The points can be joined so that the area enclosed by the lines represent the area cover of the ground ivy.

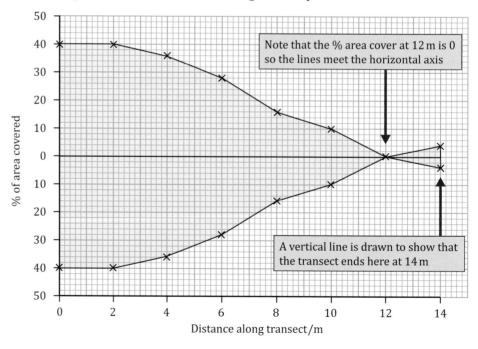

Note that the % area cover at 12 m is 0 so the lines meet the horizontal axis

A vertical line is drawn to show that the transect ends here at 14 m

Kite diagrams for each species along the transect can be drawn below each other on the same scale, to give a full description of the change in species distribution along the transect. Kite diagrams can also be drawn vertically rather than horizontally, and that is often the way you will see those showing a transect across a beach.

Plant	% cover
clover	11
daisy	14
scarlet pimpernel	5
cow wheat	5
meadow grass	65

3.7 Pie charts

Pie charts are used as an alternative to bar charts to show proportions, where the sectors of a circle are proportional to the percentage of each value. Without a scale, they are inevitably harder to read. The pie chart shows percentage cover of herbaceous plants in a grassland, derived from the data in the table:

To calculate the angle of the sector, each plant has to have its per cent cover considered as a per cent of the area of the circle.

With clover as an example, area cover = 11%.

$$\therefore \quad \text{angle of sector} = 11\% \text{ of } 360° = \frac{11}{100} \times 360 = 39.6°$$

This type of graph may be easily compiled using a spreadsheet and graph-plotting package.

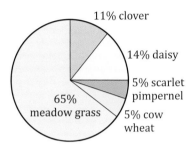

The cell cycle

Pie charts are sometimes used to display the cell cycle. Then the angle of the sector and its area are both proportional to the time spent in that phase. In the pie chart shown here, for example, the angle bounding the sector marked G2 is 170°.

So the cell spends $\frac{170}{360} = 0.47$ of its time in G2.

If the cell cycle were 24 hours long, then the cell would spend 0.47 × 24 = 11.3 hours in G2.

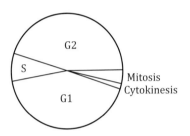

3.8 Nomograms

Nomograms are drawings comprising three scaled lines to allow computation. If two values are known, the third may be read by projecting a line joining the first two.

A simple nomogram is shown below. It relates to the digestion of protein over 50 minutes. At any time, a Biuret test may be performed using a standard method, and the absorbance following the test read in a colorimeter. From the nomogram, after 10 minutes of digestion, if the absorbance is 0.2 AU, then 0.5 g of protein has been digested.

Test yourself 3

❶ The graph shows the lengths of two populations of salmon.

What type of variation is shown by group A?

Explain your answer.

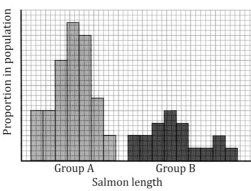

❷ The table shows the concentration of dissolved oxygen in a freshwater stream.

a) Plot the data.

b) Find the percentage of dissolved oxygen at 300 m.

Distance from sewage outflow / m	Concentration of dissolved oxygen / AU
0	9
200	12
400	18
600	30
800	39
1000	103

❸ The graph shows the change in concentration of two substances, A and B, in the cytoplasm of *Amoeba* kept in equal concentrations of A and B for 6 hours.

What is the difference in their uptake at 4 hours?

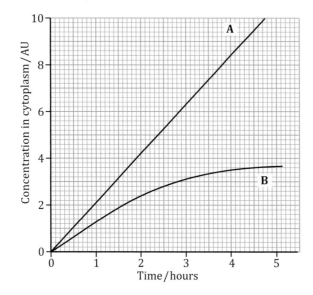

4 The number of cells in a liquid culture of plant tissue was estimated over the first six days of the culture, in either the presence or absence of the plant growth regulator cytokinin. The graph shows the number of cells per cm^3.

a) Calculate the percentage increase in cell number with cytokinin at 5 days.

b) Calculate the percentage increase in cell number at 5 days resulting from the presence of cytokinin, compared with its absence.

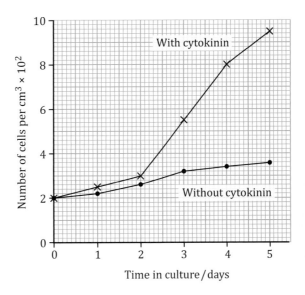

5 The graph shows the percentage loss of mass of seaweed when exposed to warm air.
Calculate the actual mass loss of 250 g seaweed after 2 hours exposure.

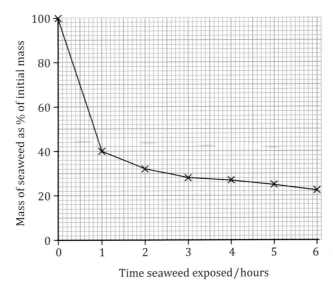

6 The graph on page 78 shows oxygen dissociation curves for human haemoglobin at two temperatures.

Using the graph:

a) What per cent oxygen is released when blood containing haemoglobin that is 90% saturated flows from a part of the body at 38 °C to one at 43 °C?

b) If blood that is 100% saturated contains 105 cm^3 oxygen / dm^3 at 38 °C, what volume of oxygen is released when blood containing haemoglobin that is 90% saturated flows from a part of the body at 38 °C to a part of the body at 43 °C?

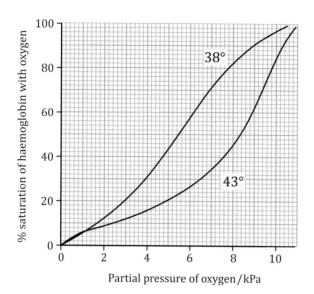

7   When red blood cells burst, they are said to undergo haemolysis. The table shows the percentage of red blood cells that undergo haemolysis in solutions of potassium chloride at different concentrations.

a)   Plot the graph.

b)   Calculate the difference in percentage of red blood cells that remain intact at 0.63 mol dm^{-3} and 0.53 mol dm^{-3}

Concentration of potassium chloride / mol dm^{-3}	% red blood cells haemolysed
0.45	98
0.50	88
0.55	78
0.60	54
0.65	28
0.70	13

Chapter 4

Scale

Living organisms are organised. That means that they obey certain rules, such as those of physics, chemistry and mathematics. A consequence of this is that in many ways, their behaviour is predictable. But the same could be said of other structures, such as the solar system or an atom. Consider the table on the right.

All these structures are organised. Living organisms are in a particular size range, in the middle of this list, which conveniently covers the whole universe. Does this mean, then, that the significant factor that separates living organisms from other structures in the universe is the scale at which they operate? That depends in part on what is meant by the term 'living organism', a point that is not lost on philosophers and science fiction fans. Sadly, it is beyond the scope of this book, which limits itself to the scale 10^{-6}–10^1 m and does not attempt to define life.

Human brains do not process very big or very small numbers well, so it is quite hard to visualise this vast range of scale. On the scale studied in Biology, if you scaled up a human to be the length of the UK, from John O'Groats to Land's End, it would be approximately a million-fold magnification. With the same magnification, an HIV would be the diameter of a golf ball.

Biology is visual. It is often the interest in looking at plants or animals that attracts people and much has been learned about structure and behaviour by observing organisms. Published material uses photographs and diagrams, but unless you know the scale, it may be hard to interpret the image.

Structure	Approximate diameter / m
Universe	10^{24}
Galaxy	10^{20}
Solar system	10^{13}
Earth	10^7
Living organisms	10^{-6}–10^1
Red blood cell	10^{-5}
Virus	10^{-7}
Molecules	10^{-9}
Atoms	10^{-10}
Atomic nucleus	10^{-15}
Sub-atomic particles	10^{-16}

If the image is this:

or this:

or this:

you intuitively know how big the organism is in real life.

But if it looks like the picture on the left, you do not know if it is the size of a pin-head or a dinner plate. A scale must therefore be added to any image, so that its true dimensions are clear. This is especially important when looking at photomicrographs taken through a light microscope or electron micrographs taken by an electron microscope because structures at these scales are less familiar. It is also important in diagrams, as the objects they represent could be any size.

4.1 Units

In Biology, the use of units is that laid down by the Institute of Biology, so SI (Système Internationale) units are used. The SI unit of length is the metre, with the abbreviation m. This system allows for bigger and smaller units indicated by prefixes. Depending on what is being studied, the scale at which Biology operates principally uses five units: the kilometre (km), the metre (m), the millimetre (mm), the micron or micrometre (μ or μm) and the nanometre (nm). Here are two ways to see how they are related:

» Pointer
Include an indication of scale on any image.

» Pointer
Always use SI units.

 4.1

a) How many mm in 20 km?
b) How many microns is 3.4 m?
c) How many nm in 1 m?
d) How many m is 3400 nm?

1 km = 1000 m	1 m = 0.001 km
1 m = 1000 mm	1 mm = 0.001 m
1 mm = 1000 μ or μm	1 μm = 0.001 mm
1 μm = 1000 nm	1 nm = 0.001 μm

Or you could relate them using powers of 10:

1 km = 10^3 m	1 m = 10^{-3} km
1 m = 10^3 mm	1 mm = 10^{-3} m
1 mm = 10^3 μ or μm	1 μm = 10^{-3} mm
1 μm = 10^3 nm	1 nm = 10^{-3} μm

4.2 Converting between units

Prefixes are used with various units. They are listed in the table below, with the factor by which they multiply. As long as you can use powers of 10, you will find it easy to interconvert them.

Prefix	Symbol	Factor
nano	n	10^{-9}
micro	μ	10^{-6}
milli	m	10^{-3}
kilo	k	10^3
mega	M	10^6
giga	G	10^9

Because of the size of living organisms, the unit cm is commonly used, even though it is not an SI unit.

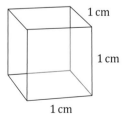

As 1 cm = 10 mm, it is easy to convert between mm³ and cm³. The cube on the left has sides of 1 cm (10 mm). Its volume is ($1 \times 1 \times 1$) = 1 cm³ or ($10 \times 10 \times 10$) = 10^3 = 1000 mm³.

So to convert cm³ into mm³, you multiply by 1000 and to convert mm³ into cm³, you divide by 1000.

Knowing the relationships between units allows them to be interconverted. As SI units all relate to each other by a small number of powers of ten, the interconversion is straightforward.

Example: an earthworm is 15 cm long and has an area of cross section of 0.5 cm². To find its volume in mm³, first find its volume in cm³ then multiply by 1000.

Volume = length × area of cross-section

= 15 × 0.5 = 7.5 cm³

= 7.5 × 1000 = 7500 mm³

>> *Pointer*
1 cm³ = 1000 mm³.

quickfire >> **4.2**

a) How many mm² in 1 cm²?
b) How many mm³ in 1 cm³?
c) How many mm³ in 10 cm³?
d) How many cm³ in 1500 mm³?

4.3 How to indicate scale

The scale of a drawing or photograph and the size of the actual object can be shown in two ways, either drawing a bar and annotating it with a distance or giving the magnification. These are both explained below.

Normally, you would choose a round number and draw a bar that long, such as shown for this buttercup. Using a round number is sensible because in any sample of the plant, the structures will be different sizes. But sometimes, you may have a diagram where a precise distance is known, such as in the diagram of the DNA molecule, where the pitch of the helix is 34 nm. Then it makes sense to give that exact measurement, as shown below.

This works for objects that you draw with the naked eye as well as those using the microscope. The only difference with the microscope is that you have to calibrate it first.

>> *Pointer*
Use a round number or a precise number when choosing the length of the scale bar.

10 mm

Buttercup

4.4 Microscope calibration

Calibrating a microscope lets you measure the actual size of structures on a microscope slide. You need an eyepiece graticule, which is the ruler you can see. It is inside the eyepiece. It is graduated 1–10 with 10 subdivisions between each number. So the eyepiece graticule has 100 eyepiece units (epu) along its length. It looks like this:

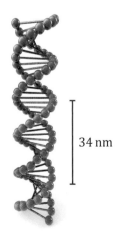

34 nm

DNA double helix

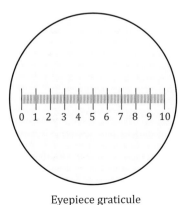

Eyepiece graticule

≫ Pointer

The eyepiece graticule looks the same in every objective but the length it represents is different.

Stage micrometer

It looks the same in every objective because it is in the eyepiece. With objectives of different magnifications, the divisions on the eyepiece graticule represent different lengths. The higher the power of the objective, the smaller length each division of the graticule represents. So calibrating has to be done for each objective.

To know what length each division represents, you need a stage micrometer. This is a microscope slide on which the object is a line 1 mm long. It is ruled with markings for tenths and hundredths of a millimetre as shown on the left. There are 100 divisions on the stage micrometer, so each stage micrometer unit (smu) = 0.01 mm or 10 μm.

To calibrate the graticule for a particular objective lens, line up graduations of the eyepiece graticule and the stage micrometer that are easy to read. Make sure the graduation lines are absolutely parallel. Look along the scales and see where they coincide again. Here are two examples:

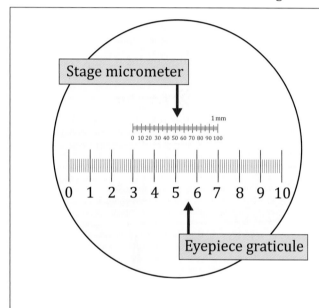

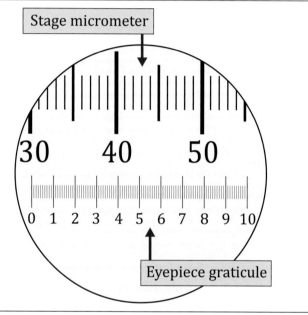

on a ×4 objective, 100 stage micrometer divisions may line up exactly with 40 eyepiece divisions	on a ×40 objective, 20 stage micrometer divisions may line up exactly with 80 eyepiece divisions

Now make three statements for the calculation, remembering to say which objective you have used:

At a ×4 objective lens	At a ×40 objective lens
1 40 epu = 100 smu	1 80 epu = 20 smu
2 1 smu = 0.01 mm	2 1 smu = 0.01 mm
3 1 epu $= \dfrac{100}{40}$ smu	3 1 epu $= \dfrac{20}{80}$ smu
$= \dfrac{100}{40} \times 0.01$ mm	$= \dfrac{20}{80} \times 0.01$ mm
= 0.025 mm	= 0.0025 mm

If other measurements are given in micrometres, the calculation could be completed like this:

$$= 0.025 \times 1000 \ \mu m$$
$$= 25 \ \mu m$$

Very small numbers given in mm are better expressed as microns so you can complete the calculation like this:

$$= 0.0025 \times 1000 \ \mu m$$
$$= 2.5 \ \mu m$$

≫ Pointer

Remember to include a key for abbreviations:

smu = stage micrometer unit;
epu = eyepiece unit.

You can work out similar calibrations for all the objectives that you use. Microscopes differ slightly in their calibration, so you have to do it again for each microscope you use. That is why, in school, it helps if you always select the same microscope. Then you only have to do it once. We will see how to use the calibration in measuring when we discuss biological drawings.

a) For a macroscopic specimen, i.e. a specimen you can see without artificial magnification

Let us imagine you have calibrated your microscope; you are using a ×4 objective lens to examine a transverse section of a stem and have made a diagram of it. Here is how to work out how to do the scale bar.

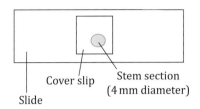

Diagram of slide with stem section

You could measure the diameter of the stem on the actual slide and if it were, say 4 mm diameter, you could choose a suitable round number, such as 1 mm. This will be the length of the scale bar. Now measure the diameter of your diagram. Perhaps it is 160 mm diameter. This 160 mm is equivalent to the 4 mm of the actual stem. So 1 mm of the actual stem = 160 ÷ 4 = 40 mm on the diagram. You could draw a bar 40 mm long and label it 1 mm, as shown on the right.

b) For a microscopic specimen

You may wish to use a scale bar but be unable to directly measure the size of your specimen as it is microscopic. In that case you have to use the calibration on the microscope. Let us consider the nucleus of a cell, another object that might appear circular in the diagram. Let us say that you are using a ×40 objective lens, and that with this objective, the diameter of the nucleus is 6 eyepiece units, as measured on the eyepiece graticule. You can use the following steps to calculate its actual diameter:

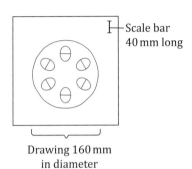

Diagram of drawing with scale bar

> With a ×40 objective, using the eyepiece graticule,
>
> $$\text{nucleus diameter} = 6 \text{ epu}$$
>
> from the calibration, $1 \text{ epu} = 2.5\,\mu$
>
> $$6 \text{ epu} = 6 \times 2.5\,\mu = 15\,\mu m$$
>
> i.e. the nucleus has a diameter of 15 μm.

Let us say that you decide a 5 μm scale bar would be suitable. If the diagram you have drawn has a nucleus with a diameter 30 mm, this is how to work out the length of the scale bar:

> $$15\,\mu m \text{ on slide} \equiv 30 \text{ mm on diagram}$$
>
> $$\therefore 1\,\mu m \text{ on slide} \equiv \frac{30}{15}\,\text{mm on diagram}$$
>
> $$\therefore 5\,\mu m \text{ on slide} \equiv \frac{30}{15} \times 5 = 10 \text{ mm on diagram}$$
>
> You draw a bar 10 mm long and label it 5 μm.

4.5 Magnification

Magnification tells you the size of the image in relation to the object. It is calculated by the equation

$$\text{Magnification} = \frac{\text{Image size}}{\text{Object size}}$$

If an image has a magnification greater than 1, it is larger than the actual object, and if its magnification is less than one, it is smaller. So if an image is labelled ×2 it means that it is twice the size of the actual object.

If it is labelled × $\frac{1}{2}$ it is half the size. If it is labelled ×1 it is life-sized.

The photographs on the left show an African elephant and a cabbage white butterfly, apparently the same height. Your general knowledge tells you that the butterfly is, in fact, smaller than the elephant, but the magnifications given confirm this and let you infer their actual sizes.

Given any two of these three factors in the equation, it is possible to calculate the third by rearranging:

$$\text{Object size} = \frac{\text{Image size}}{\text{Magnification}} \quad \text{and} \quad \text{Image size} = \text{Object size} \times \text{Magnification}.$$

When you make these calculations, it will help you to think clearly if you write the information that you have and write the equation. Check your units. If you are comparing the sizes of two organisms, the units must be the same.

$\times \frac{1}{100}$

An elephant

$\times 2$

A cabbage white butterfly

For the elephant:	For the butterfly's body length:
Magnification $= \times \dfrac{1}{100}$	Magnification $= \times 2$
Image height $= 30\,\text{mm}$	Image length $= 30\,\text{mm}$
Object height $= \dfrac{\text{image height}}{\text{magnification}}$	Object length $= \dfrac{\text{image length}}{\text{magnification}}$
$= 30 \div \dfrac{1}{100}$	$= \dfrac{30}{2}$
$= 30 \times 100$	$= 15\,\text{mm}$
$= 3000\,\text{mm}$	
As this is an elephant, it makes sense to convert mm into m:	
$1\,\text{m} = 1000\,\text{mm}$	
$\therefore\ 3000\,\text{mm} = 3\,\text{m}$	

The same principle works for microscope images, and a frequent task in an examination is to calculate the actual size of cells or organelles.

The photograph shows the cells of the lower epidermis of a leaf, studded with stomata. The photomicrograph has a magnification of × 800. Calculate the length of the epithelial cell marked A.

> Magnification $= \times 800$
>
> Length of image of cell $= 40\,\text{mm}$
>
> Actual cell length $= \dfrac{40}{800}\,\text{mm} = 0.05\,\text{mm}$
>
> As this is a cell, micrometres are a more appropriate unit.
>
> Length $= 0.05\,\text{mm} = 0.05 \times 1000\,\mu\text{m} = 50\,\mu\text{m}$

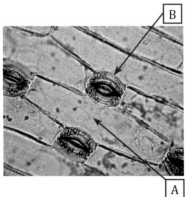

B

A

Lower epidermis of a leaf

quickfire ⟫ 4.3

a) Calculate the length of the elephant's trunk.

b) Calculate the maximum width of the butterfly's body.

quickfire ⟫ 4.4

Calculate the maximum width of stoma B.

Some people like to use the triangle method to work out how to rearrange the equation. Here is the triangle:

It is easy to see:

a) $I = M \times O$

b) $M = \dfrac{I}{O}$

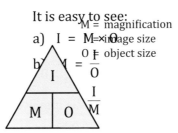

M = magnification
I = image size
O = object size

4.6 Area

You may need to calculate an area; if, for example, you are doing practical ecology and need to know the area of a quadrat or a field. A common technique is to approximate the area to a square or a rectangle and then multiply the lengths of two sides together:

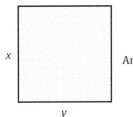

$\text{Area} = (x \times y)$

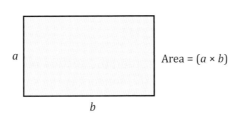

$\text{Area} = (a \times b)$

If the units of the lengths of the sides are metres, the units of area are m^2.

For example, if $x = 10$ m and $y = 15$ m, area = $(10 \times 15) = 150$ m^2. But other units of area are used in Biology. As with the linear measurements, there are two ways to see how units of area are related:

1 km^2 = 1000 × 1000 = 1 000 000 m^2
1 m^2 = 100 × 100 = 10 000 cm^2
1 m^2 = 1000 × 1000 = 1 000 000 mm^2
1 mm^2 = 1000 × 1000 = 1 000 000 μm^2

1 m^2 = 0.000001 km^2
1 cm^2 = 0.0001 m^2
1 mm^2 = 0.000001 m^2
1 μm^2 = 0.000001 mm^2

Or you can relate them using powers of ten:

1 km^2 = 10^6 m^2
1 m^2 = 10^4 cm^2
1 m^2 = 10^6 mm^2
1 mm^2 = 10^6 μm^2

1 m^2 = 10^{-6} km^2
1 cm^2 = 10^{-4} m^2
1 mm^2 = 10^{-6} m^2
1 μm^2 = 10^{-6} mm^2

Once you have done your calculation, you may find the units are not appropriate if the numbers are very large or very small. In that case, you need to convert your unit into another. Here is an example:

A square frame 10 × 10 gridded quadrat has sides of length 0.5 m. What is the area of each square in the grid?

The total area = 0.5 × 0.5 = 0.25 m².

There are 10 × 10 = 100 squares in the grid

∴ each square has an area = $\frac{0.25}{100}$ = 0.0025 m².

You could convert this to mm² as mm are SI units. 1 m = 1000 mm

∴ 1 m² = 1000 × 1000 = 1 000 000 mm²

∴ each square has an area = 0.0025 × 1 000 000 = 2500 mm².

This is also quite an awkward number for the human brain. The intermediate unit, cm³, would be more suitable, even though it is not an SI unit. But as it is suitable, it is reasonable to use it.

1 m = 100 cm ∴ 1 m² = 100 × 100 = 10 000 cm²

∴ each square has an area = 0.0025 × 10 000 = 25 cm².

 4.6

Find the following areas and put them into more suitable units:

a) A square quadrat with sides of 500 mm.

b) A rectangular area of desert, with sides of 450 m and 1050 m.

c) A square sample of a leaf, with sides 30 mm.

4.7 Volume

If you have understood the section on area, you will understand the section on volume, because the concept is the same. All that happens is you add another dimension, so whereas, in 2D, area = length × height, in 3D, volume = length × height × depth:

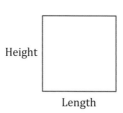
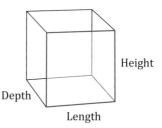

》 Pointer
Volume = length × height × depth.

As before, there are two ways to see how units of area are related:

1 km³ = 1000 × 1000 × 1000 = 1 000 000 000 m³	1 m³ = 0.000000001 km³
1 m³ = 100 × 100 × 100 = 1 000 000 cm³	1 cm³ = 0.000001 m³
1 m³ = 1000 × 1000 × 1000 = 1 000 000 000 mm³	1 mm³ = 0.000000001 m³
1 mm³ = 1000 × 1000 × 1000 = 1 000 000 000 µm³	1 µm³ = 0.000000001 mm³

Or you can relate them using powers of ten:

1 km³ = 10^9 m³	1 m³ = 10^{-9} km³
1 m³ = 10^6 cm³	1 cm³ = 10^{-6} m³
1 m³ = 10^9 mm³	1 mm³ = 10^{-9} m³
1 mm³ = 10^9 µm³	1 µm³ = 10^{-9} mm³

The unit km³ is useful when thinking about oceans, such as in the study of fish stocks. You may come across mm³ when learning about the number of blood cells in a given volume of blood. When thinking about cells, µm³ may be useful.

When measuring volumes for biochemistry experiments, the most frequently used unit is cm^3, even though cm is not an SI unit. Numerically, the units are the same as ml (millilitres), i.e. $1\ cm^3 = 1\ ml$. But ml is also is not an SI unit. The current convention, though, is to use cm^3 as it is more closely related to SI units than ml.

Similarly, dm^3 is the preferred larger unit of volume, rather than litres, although this is sometimes used.

Remember, there is no unit ml^3. That would have 9 dimensions and does not exist in the universe as we understand it.

A word of warning

It is useful to estimate the answer to a calculation before you start. That way you will know if the answer is sensible. Methods of estimation were discussed on page 15.

When you do a calculation you should always show your working. This means that you present a series of arithmetical statements which are equations. To keep it simple, on each line, have only one equation and one equals sign. Make sure that what is on the right of the equals sign is really equal to what is on the left.

The danger is in thinking faster than you write, so that you miss out some of the steps in the logic. Often, this results in powers of ten or units being wrong. Take care to include every step, and check your working at the end.

If it does not correspond with your estimate, think again. One, or both, of them may be wrong.

>> Pointer

Always use units that give you a number you can visualise.

quickfire 4.7

Find the volumes of these structures and decide on the most suitable units:

a) A cuboidal epithelial cell in the nephron which has sides of 0.03 mm.

b) An experimental fish tank with sides 1.5 m × 0.6 m × 0.9 m.

4.8 Constructing ecological pyramids

Ecological pyramids show the number, biomass or energy at different trophic levels. In each pyramid, the area of the bar is proportional to the number, biomass or energy at that trophic level. Following an investigation into the structure of a habitat, you may wish to construct a pyramid. The principles are the same for each type of pyramid, and the method will be illustrated using biomass data.

If invertebrates in a freshwater pond were being investigated, animals in kick samples from the benthic layer would be identified, counted and attributed to trophic levels, using published information. The average mass of one individual could be found from published material and the total mass of each species calculated. Then the total mass in each trophic level would be calculated, producing a table like this:

Organism	Mean number / quadrat	Mass of one individual / g	Total mass / mg	Feeding type	Trophic level	Total mass at trophic level / g
Leech	4	1.0	4.0	Primary consumer	2	30.8
Flatworms	16	1.0	16.0			
Beetle larva	6	1.8	10.8			
Alderfly larva	1	9.0	9.0	Secondary consumer	3	11.2
Damselfly nymph	1	2.2	2.2			
Biting midge larva	2	0.3	0.6	Decomposers	D	11.0
Water shrimp	4	2.6	10.4			

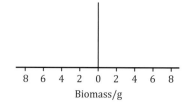

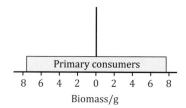

In the pyramid, each bar has the same height and so its length is proportional to the mass at that trophic level. The pyramid is symmetrical around a vertical axis in the centre of each bar.

The largest biomass in this set of data is 30.8 g. Given the size of the graph paper, the scale 2 g : 1 cm could be used. A vertical line is drawn to represent the centre of the pyramid and half the biomass will be drawn on each side. A horizontal line is drawn to represent the biomass axis:

The primary consumers have a biomass of 30.8 g.

$2\,g \equiv 1\,cm \therefore 1\,g \equiv 0.5\,cm$

$\therefore 30.8\,g \equiv 30.8 \times 0.5\,cm = 15.4\,cm$

So the bar representing the primary consumers must be 15.4 cm long. Half of this, 7.7 cm is drawn either side of the vertical axis and the bar is labelled:

The secondary consumers have a biomass of 11.2 g.

$2\,g \equiv 1\,cm \therefore 1\,g \equiv 0.5\,cm$

$\therefore 11.2\,g \equiv 11.2 \times 0.5\,cm = 5.6\,cm$

So the bar representing the secondary consumers must be 5.6 cm long. Half of this, 2.8 cm is drawn either side of the vertical axis and the bar is labelled:

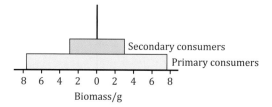

All trophic levels feed into the group called decomposers. Placing the decomposers directly above the secondary consumers would suggest that only they feed into the decomposers. Instead, decomposers can be represented by a bar of the appropriate length, drawn at an angle, as shown here.

The decomposers have a biomass of 11.0 g.

$2\,g \equiv 1\,cm \therefore 1\,g \equiv 0.5\,cm$

$\therefore 11.0\,g \equiv 11.0 \times 0.5\,cm = 5.5\,cm$

So the bar representing the secondary consumers must be 5.5 cm long. Half of this, 2.75 cm is drawn either side of the vertical axis and the bar is labelled:

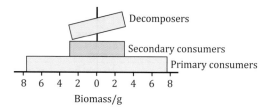

Sometimes, instead of drawing a bar at an angle, the decomposers are shown as a vertical box at the side of the pyramid, adjacent to all the other trophic levels, as shown here:

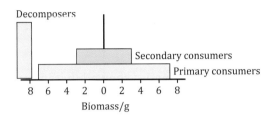

Test yourself   4

1 On aligning an eyepiece graticule with a stage micrometer, it was observed that, using a ×4 objective lens, 20 stage micrometer divisions lined up with 6 eyepiece units. Calculate the length of each eyepiece unit using this objective lens.

2 On aligning an eyepiece graticule with a stage micrometer, it was observed that, using a ×10 objective lens, 40 stage micrometer divisions lined up with 45 eyepiece units. Calculate the length of each eyepiece unit using this objective lens.

3 A lymphocyte is approximately spherical and has a diameter of 20 μm. The volume of a sphere is $\frac{4}{3}\pi r^3$.

Calculate the volume of a lymphocyte.

4 Here is a diagram showing part of a cell membrane. In the drawing, it has been magnified 3 million times.

Calculate the actual width of the cell membrane in μm.

Show your working.

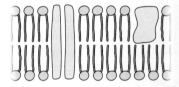

5 The diagram shows the structure of a leaf cell from the palisade layer. The maximum length of the vacuole is 40 μm.

Calculate the magnification of the drawing. Show all your working.

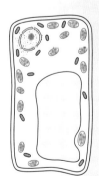

6 The liver cell below has been magnified 1000 times.

Calculate the actual diameter of its nucleus.

Show your working.

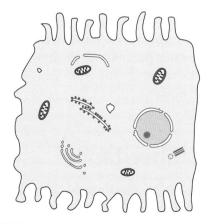

Chapter 5

Ratios and their use in genetics

In the 19th century, experiments in genetics showed that in some crosses, the proportions of phenotypes in the offspring could be predicted. In fact, Gregor Mendel's counts were so close to his predicted outcomes that the numbers he reported seem statistically too good to be true. No-one would expect an Augustinian monk in the 1850s to lie about his results. Maybe, to save Mendel being disappointed by not having perfect answers, and to save him the embarrassment of having his theories unsupported by observations, the young monks who worked with him by counting plants in the greenhouses, massaged the results to make them fit the predictions almost perfectly. It was too perfect because the young monks did not know any statistics.

But this is not the earliest record of the application of genetic ratios. We are told the biblical story of Jacob, who was to be given all the speckled sheep and goats in Laban's flocks as wages. Jacob cared for the flocks and chose which animals would be allowed to mate. There are many ways to interpret this story, and one way is in terms of Mendelian genetics. In this explanation, the gene for being speckled was dominant over that for being monochrome. Heterozygous speckled animals came on heat earlier than the homozygous speckled animals and had more offspring. They were more vigorous and so they could easily be identified. Jacob, obviously a proto-geneticist, put these heterozygous animals together to mate. They would produce 75% speckled offspring, of which two thirds would be as vigorous as their parents, being heterozygous also. The weaker, homozygous speckled animals would mate later and would produce fewer offspring. These offspring would all be speckled, but as weak as their parents. Monochrome animals were homozygous recessive and would produce only monochrome offspring. To reduce Jacob's wages Laban, who had not done genetics at school, secretly tried to remove the speckled offspring. But Jacob knew this and secretly made the weaker (homozygous) speckled animals available to Laban, he would have been able to increase the proportion of the flock with vigorous, (heterozygous) speckled animals, increasing his wages by his understanding of genetics.

So understanding genetic ratios can help you grow the plant types you want, can increase your wages and can improve your A Level grade.

Speckled sheep

》 Pointer
Here are the ratios you have to recognise:
$1:0, 1:1, 3:1, 1:2:1, 9:3:3:1$ and $1:1:1:1$.

There are several ratios that you are required to recognise. If you are given the proportions of phenotypes in a population, you should immediately be able to identify the ratio and know what genetic cross would have produced it. Here are the ratios you need to know: $1:0, 1:1, 3:1, 1:2:1, 9:3:3:1$ and $1:1:1:1$. These will all be explained.

5.1 Blending inheritance

Before genes were understood, it was thought that offspring had characteristics of both parents because the blood of the two parents was mixed and blood was the source of the characteristics. Parental characteristics were blended in the offspring. This ignored an important implication. If every generation were an average of the previous one, after a few generations of interbreeding, we would all be almost identical. We are not.

5.2 Monohybrid inheritance

Sexually reproducing organisms have many thousands of genes. When geneticists discuss monohybrid inheritance, they are considering only one gene and ignoring the rest. A genetic definition of a gene could be a sequence of DNA that codes for a characteristic. So we equate one gene with one characteristic. It was the consideration of one gene and only one characteristic which showed the idea of blending inheritance to be false. A characteristic could disappear in one generation only to reoccur in the next. So it could not have been blended. One of Mendel's great intellectual leaps was to recognise this and as a result, he developed the idea of particulate inheritance. The characteristics, in this view, were identified with particles which were passed down the generations. Sometimes they were expressed and sometimes hidden, only to reappear in a subsequent generation. Now we call those particles 'genes'.

5.2.1 Crosses and offspring

Mendel's famous crosses produced particular ratios of characteristics in the offspring which are referred to as 'Mendelian' or 'ideal' ratios. But the numbers in the offspring are rarely exactly equal to the predictions. For example, although the chances of having a baby boy or girl are equal, many two-child families do not have one of each. The ratios predicted by Mendel are ideal in the sense that you would see them if certain assumptions are true. These are that gametes and zygotes behave perfectly, and there are no other factors involved.

It is assumed that:

a) All males and females mate at random.

b) Gamete types are formed in equal numbers.

c) All gametes can all fuse with each other with equal probability.

d) All zygotes have an equal chance of survival.

If these assumptions were true, the ideal ratio may still not be seen because it is a matter of chance which egg and which sperm fuse together. The more offspring there are, the more the ratio will approach the ideal, which is why genetics experiments need to produce a large number of offspring. That is possible when dealing with fungi and many plants, but not if, for example, you are looking at the human children of human parents. Even then, the numbers will still not be exactly what is predicted and so a statistical analysis must be done. The statistics will be explained in Chapter 7. This chapter is just about ratios.

5.2.2 Monohybrid crosses

Two homozygous parents with the dominant characteristic (TT × TT)

A pure-breeding plant is homozygous and can only produce one type of gamete. If two pure-breeding tall pea plants are crossed, each produces the same gamete type, so all the offspring are identical:

A pea plant

quickfire 5.1

The allele for red eyes in fruit flies is dominant over the allele for white eyes. If homozygous red-eyed flies mate with homozygous white-eyed flies, what will be the ratio of phenotypes in the F_1?

quickfire 5.2

In guinea pigs, the allele for a rough coat is dominant over the allele for a smooth coat. Describe the progeny of a cross between two smooth-coated guinea pigs.

»» Pointer

When you quote a genetic ratio, make it very clear which number represents which phenotype.

»» Pointer

TT × TT gives the ratio 1 : 0 tall : dwarf.

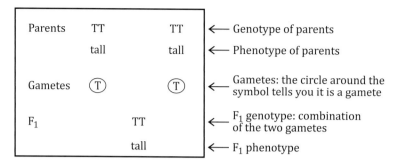

Only tall plants are produced because neither parent has an allele for the dwarf characteristic. The phenotypic ratio tall : dwarf = 1 : 0

So if a cross is described where all the offspring have the same phenotype, it could be because both parents are pure-breeding and have the same alleles.

Two homozygous parents, one with the dominant and one with the recessive characteristic (TT × tt)

Another way to get identical offspring is to cross two homozygous plants, where one has dominant alleles and the other has recessive alleles. One parent only produces gametes containing the dominant allele and the other only produces gametes with the recessive allele. So all offspring have one dominant and one recessive allele. They are heterozygous but have the dominant characteristic. The recessive characteristic is not expressed as its allele is not homozygous.

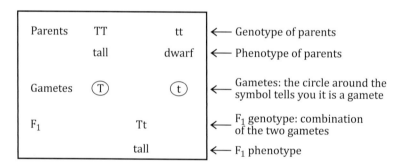

»» Pointer

TT × tt gives the ratio 1 : 0 tall : dwarf.

One homozygous dominant parent and one heterozygous parent (TT × Tt)

The homozygous parent can only produce one type of gamete. The heterozygous parent produces two types of gamete in equal proportions. The Punnett square shows their offspring.

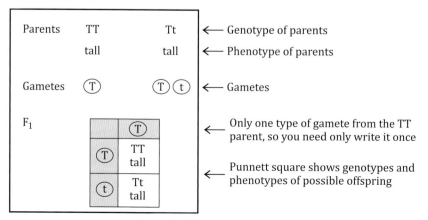

»» Pointer

When you draw a Punnett square, leave enough room in each cell of the table for genotype and phenotype.

All the offspring have a dominant allele (T) so all of them are tall, even though they may have either the dominant or recessive allele from the other parent. So the phenotypic ratio from this cross is 1 : 0 tall : dwarf.

One homozygous recessive parent and one heterozygous parent (tt × Tt)

One parent can only make gametes with a recessive allele. The other parent can make two types of gamete and so the progeny will not all be the same.

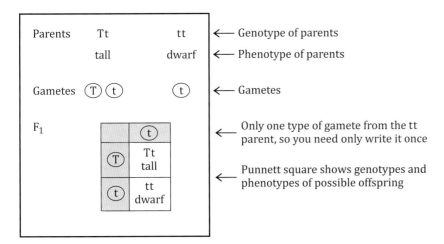

Parents — Genotype of parents
— Phenotype of parents
Gametes — Gametes
Only one type of gamete from the tt parent, so you need only write it once
Punnett square shows genotypes and phenotypes of possible offspring

This gives a phenotypic ratio 1 : 1 tall : dwarf. So each individual has a 50 : 50 chance of being either tall or dwarf. Another way of saying this is that each offspring has a 0.5 probability, or 50% chance, of being tall or dwarf. It does not mean that if there were two offspring one would be tall and one dwarf. They could be, but each one, individually, has a 50% chance of being either tall or dwarf.

The test cross: If you had a tall plant, it could be either TT or Tt. The way to find out which it is, is to cross it with a homozygous recessive, tt, plant. The proportions in the F_1 tell you what the tall parent's genotype was. If the plant were TT, the cross would be TT × tt and all the offspring would be Tt and therefore tall. But if the tall plant were Tt, the cross would be Tt × tt. The ratio in the offspring would be 1 : 1 tall : dwarf. So by looking at the F_1 ratio, you know what the parent would have been. Crossing with tt has displayed all the alleles of the unknown parent in the F_1. You have been able to test the unknown parent's genotype and so crossing with the homozygous recessive is called a test cross.

Two heterozygous parents (Tt × Tt)

If an individual is heterozygous, it can make gametes containing either of the alleles. So there are two types of gamete. When meiosis occurs, the alleles are separated into different daughter cells. Gametes containing the two different alleles are produced in equal numbers. Another way of saying this is that the probability of having either allele in a gamete is 0.5. Both parents make two types of gamete and the way that they can combine is illustrated by the Punnett square.

>> *Pointer*
TT × Tt gives the ratio 1 : 0 tall : dwarf.

>> *Pointer*
Tt × tt gives a 1 : 1 ratio.

quickfire >> 5.3

The allele for normal wing length in *Drosophila*, the fruit fly, is dominant over the allele for short, or vestigial, wings. What are the genotypes of the flies that mate to produce the ratio 1 : 1 normal wing length : vestigial wings in their offspring?

>> *Pointer*
When more than one type of gamete is made by a parent, a Punnett square is the best way of finding out the genotypes of the offspring.

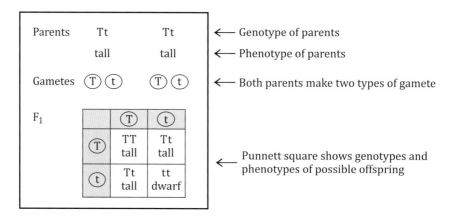

In the F_1, the TT is tall as it has only dominant alleles. The Tt individuals are tall as they have the dominant allele. The tt is dwarf as it only has recessive alleles. This gives a phenotypic ratio of 3 : 1 tall : dwarf.

Another way of saying this is that on average, three-quarters of the offspring will be tall and one-quarter will be dwarf. Remember that is unlikely to be seen unless offspring are produced in big enough numbers to make the sample size statistically significant. So a better way of explaining the ratio would be to say that any one of the offspring has a 75% chance of being tall and a 25% chance of being dwarf. If this were humans being described, you could say that every time a woman was pregnant, the child would have a 0.75 probability of having the dominant characteristic and 0.25 probability of having the recessive characteristic.

5.2.3 Partial dominance

This refers to a situation where a gene does not have clearly defined dominant and recessive alleles.

i) Incomplete dominance

Sometimes the distinction between the dominant and recessive nature of alleles is not as clear-cut as in the case of tall and dwarf pea plants. A commonly cited example of this is in the garden flower *Antirrhinum majus*, the snapdragon. This has been bred in a variety of colours, including pure-breeding red and white flowers. These are homozygous. The allele producing red flowers is given the allele symbol R so a red flower can be designated RR. Similarly, the allele for producing white flowers has the symbol W, so the white flower can be shown as WW. Each produces one gamete type so all the offspring of a self-fertilised plant will be the same colour as the parent.

If one parent has red flowers and the other has white, and if one of these characteristics were dominant, the F_1 would be all one colour, either red or white. But the F_1 are all pink. So neither R nor W is dominant and a new phenotype is produced in the heterozygote:

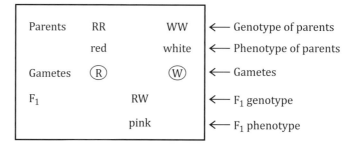

>> *Pointer*

Tt × Tt gives the ratio 3 : 1 tall : dwarf.

quickfire >> 5.4

In a cross between fruit flies with red and white eyes, all the F_1 had red eyes. What would be the phenotypic ratio in the F_2 if the red eyed F_1 interbred?

quickfire >> 5.5

In a cross between two black labradors, six of the puppies were black and two were golden. What can you deduce about the genotypes of the black parents?

Antirrhinum majus

>> *Pointer*

Where alleles show incomplete dominance, RR × WW gives the ratio 0 : 1 : 0.

We could describe this ratio as 0 : 1 : 0 red : pink : white.

When you cross two F_1 heterozygotes, the normal, Mendelian phenotypic ratio in the F_2 is 3 : 1, as described above for Tt × Tt. But with incomplete dominance, the outcome is different because RW has a different phenotype from RR and WW:

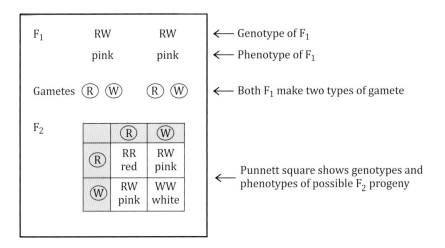

Genotype of F_1
Phenotype of F_1
Both F_1 make two types of gamete
Punnett square shows genotypes and phenotypes of possible F_2 progeny

» Pointer
RW × RW gives ratio 1 : 2 : 1.

quickfire 5.6

If a red cow (RR) mates with a white bull (WW), their calves are roan (RW). What is the phenotypic ratio in the offspring produced by two roans which mate?

Heterozygous offspring are pink and so there are three genotypes and therefore three phenotypes in the F_2. This gives a ratio 1 : 2 : 1 red : pink : white.

ii) Co-dominance

Alleles which show co-dominance are expressed in the heterozygote and produce a phenotype in which both alleles are evident. In genetic crosses, they behave in a similar way to those which show incomplete dominance, as has already been described. Co-dominance is shown in a gene associated with haemoglobin production, a mutation in which can produce sickle cell disease. The allele Hb^A codes for normal, adult haemoglobin. The mutant allele, Hb^S, produces an altered β-globin polypeptide chain. People with the homozygous $Hb^S Hb^S$ phenotype have sickle cell disease and people who are heterozygous, $Hb^A Hb^S$, have sickle cell trait. Both normal and altered β-globin are made and the symptoms are generally less severe than sickle cell disease, in which only altered β-globin is made. A cross between heterozygous people produces the standard 1 : 2 : 1 ratio:

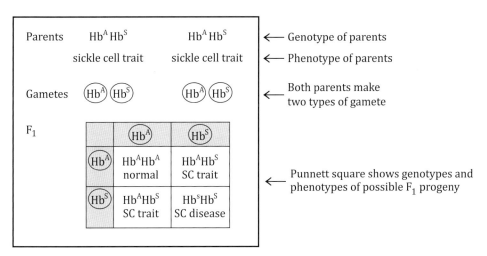

Genotype of parents
Phenotype of parents
Both parents make two types of gamete
Punnett square shows genotypes and phenotypes of possible F_1 progeny

The ratio in the progeny is 1 normal : 2 sickle cell trait : 1 sickle cell disease.

Summary so far

Cross	Monohybrid ratios
TT × TT	1 : 0
TT × tt	1 : 0
TT × Tt	1 : 0
Tt × tt	1 : 1
Tt × Tt	3 : 1
RW × RW	1 : 2 : 1

5.3 Dihybrid inheritance

When geneticists consider dihybrid crosses, they are ignoring most of the genes and concentrating on the inheritance of only two. They study two genes which, being on different chromosomes, behave independently of each other at meiosis. Geneticists investigate how the alleles of the two genes are expressed in the same offspring. One of Mendel's classic experiments looked at the simultaneous inheritance of the gene for seed colour and the gene for seed shape in peas.

In monohybrid inheritance, each gamete contained one copy of the gene of interest and so one allele appeared in the circle representing the gamete. In dihybrid inheritance, there are two genes of interest and so each gamete will have an allele of each. So two alleles will appear in the circle representing the gamete.

When you write genetic crosses, you should always define the allele symbols:

The symbol for the allele for yellow seeds is Y; the symbol for the allele for green seeds is y.

The symbol for the allele for round seeds is R; the symbol for the allele for wrinkled seeds is r.

A pure-breeding plant that always produces offspring with round, yellow seeds when it is self-fertilised is homozygous at the shape and colour genes – RRYY. Its gametes contain one allele for each gene so they are written $\textcircled{RY}$.

A pure-breeding plant that always produces offspring with wrinkled, green seeds when it is self-fertilised is also homozygous at the shape and colour genes – rryy. Its gametes contain one allele for each gene so they are written $\textcircled{ry}$. A cross between two such plants can be written using the same protocol as for monohybrid inheritance:

quickfire 5.7

How would you write the gametes for a plant where two genes are considered and the plant's genotype is AABB?

quickfire 5.8

How would you write the gametes for a plant where two genes are considered and the plant's genotype is aabb?

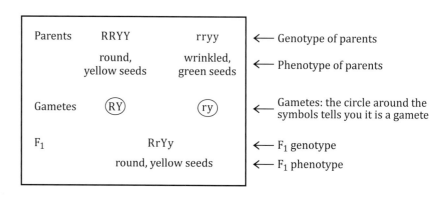

Parents	RRYY	rryy	← Genotype of parents
	round, yellow seeds	wrinkled, green seeds	← Phenotype of parents
Gametes	$\textcircled{RY}$	$\textcircled{ry}$	← Gametes: the circle around the symbols tells you it is a gamete
F$_1$	RrYy		← F$_1$ genotype
	round, yellow seeds		← F$_1$ phenotype

Here again is a 1 : 0 ratio but as there are four possible combinations, it could be described as a 1 : 0 : 0 : 0 ratio. This takes into account the four phenotypes – round yellow, round green, wrinkled yellow and wrinkled green.

Crossing F1 individuals, which are heterozygous at both genes

This F_1 RrYy is heterozygous at both genes. When it makes gametes, independent assortment in meiosis means that the R will combine equally frequently with Y and y so that the gametes RY and Ry are formed in equal proportions. Similarly, the r can combine equally with Y or y so that the gametes rY and ry are formed with the same frequency. R and r occur in equal frequency so all four gamete types are made in equal proportions. The Punnett square for deriving the genotypes of the F_2 is consequently a 4 × 4 grid.

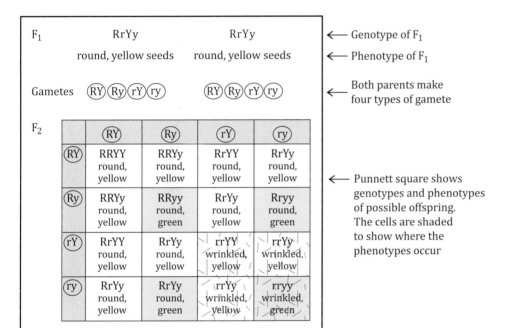

The phenotypic ratio is 9 : 3 : 3 : 1 round yellow : round green : wrinkled yellow : wrinkled green.

This is the classic 'dihybrid ratio' and when progeny are in this ratio it implies that the two parents were both heterozygous at both genes.

A thought about monohybrid and dihybrid inheritance

Here is a standard F_2 dihybrid ratio:

round, yellow	:	round green	:	wrinkled yellow	:	wrinkled green
9	:	3	:	3	:	1

If you consider only one gene at a time, the ratio yellow : green = 12 : 4 = 3 : 1.

The ratio round : wrinkled = 12 : 4 = 3 : 1.

This shows that even though the original analysis was of two genes inherited together, the system still behaves as expected when one gene on its own is considered.

> **Pointer**
RRYY × rryy gives the ratio
1 : 0 : 0 : 0.

 5.9

In tomato plants, the allele for hairy stems is dominant over the allele for smooth stems. The allele for red fruit is dominant over the allele for yellow. In a dihybrid cross between a plant homozygous dominant at both alleles and one homozygous recessive at both alleles, what will the F_1 look like?

> **Pointer**
RrYy × RrYy gives the ratio
9 : 3 : 3 : 1.

quickfire 5.10

What will be the phenotypic ratio in the F_2 of the cross in QF 5.9?

Crossing with the double homozygous recessive – the dihybrid test cross

A plant that produces round, yellow seeds could have the genotype RRYY or RRYy or RrYY or RrYy. To find out which, the plant would be crossed with one that is homozygous recessive at both genes, rryy. This is directly analogous to the test cross described for monohybrid inheritance on page 93. The rryy individual can only make (ry) gametes. The characteristics expressed in the next generation will be associated with the alleles of the gametes the (ry) gametes fuse with; so the proportions of phenotypes in the progeny will indicate the genotype of the parent producing the round yellow seeds.

The four crosses are shown below, to show how the different phenotypic proportions occur.

(i) RRYY × rryy

Yellow pea seeds

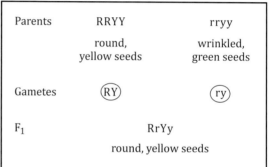

Parents	RRYY	rryy
	round, yellow seeds	wrinkled, green seeds
Gametes	(RY)	(ry)
F₁	RrYy	
	round, yellow seeds	

Green pea seeds

round, yellow	:	round green	:	wrinkled yellow	:	wrinkled green
1	:	0	:	0	:	0

(ii) RRYy × rryy

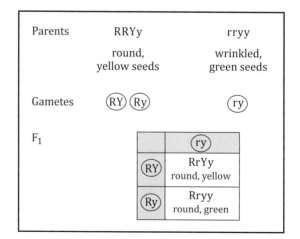

Parents	RRYy	rryy
	round, yellow seeds	wrinkled, green seeds
Gametes	(RY) (Ry)	(ry)

F₁

	(ry)
(RY)	RrYy round, yellow
(Ry)	Rryy round, green

round, yellow	:	round green	:	wrinkled yellow	:	wrinkled green
1	:	1	:	0	:	0

(iii)   RrYY × rryy

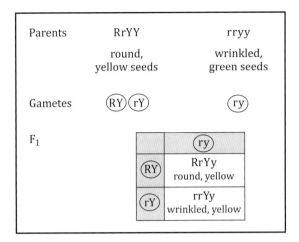

round, yellow	:	round green	:	wrinkled yellow	:	wrinkled green
1	:	0	:	1	:	0

quickfire 5.11

How would you write the gametes for a plant where two genes are considered and the plant's genotype is AaBB?

(iv)   RrYy × rryy

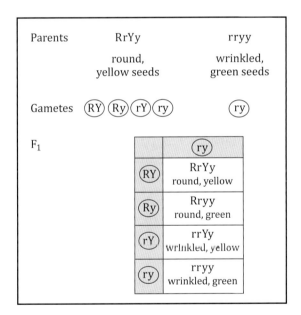

round, yellow	:	round green	:	wrinkled yellow	:	wrinkled green
1	:	1	:	1	:	1

quickfire 5.12

How would you write the gametes for a plant where two genes are considered and the plant's genotype is AaBb?

These crosses show that the proportions of phenotypes in the next generation can be used to determine the genotype of the unknown parent. The only one of these you would be expected to recognise is 1 : 1 : 1 : 1 produced by the double heterozygote, RrYy, crossed with the double homozygous recessive, rryy. If, however, you were writing an essay about this sort of thing, explaining the origin of the other ratios may be useful.

>> *Pointer*

RrYy × rryy gives the ratio 1 : 1 : 1 : 1.

Summary of crosses

The crosses shown in the table on page 100 all produce Mendelian, or ideal, ratios. You are expected to recognise these ratios and to know what crosses produced them.

Cross	Phenotypic ratios	Cross type
TT × TT	1 : 0	Monohybrid
TT × tt	1 : 0	
TT × Tt	1 : 0	
Tt × Tt	3 : 1	
Tt × tt	1 : 1	Monohybrid test cross
RW × RW	1 : 2 : 1	Incomplete dominance
		Co-dominance
RrYy × RrYy	9 : 3 : 3 : 1	Dihybrid
RrYy × rryy	1 : 1 : 1 : 1	Dihybrid test cross

Ratios and numbers of offspring

A ratio of 3 : 1 shows the chance of any one of the offspring having either of the characteristics determined by the two alleles. If there are four offspring, each has a 3 : 1 chance of having either phenotype. In many crosses, there are more than four offspring, so if you looked at, for example, 400 offspring, you might expect to see the same 3 : 1 ratio. To work out what you would expect out of 400, you add the numbers in the ratio (3 + 1) and divide that into the total number of offspring to find out how many in the class described as '1': 3 + 1 = 4

$$400 \div 4 = 100$$

So the class of 1 has 100 and the class of 3 has 100 × 3 = 300.

Another way to think about this is to say that a 3 : 1 ratio is that same as $\frac{3}{4} : \frac{1}{4}$. Then $\frac{1}{4} \times 400 = 100$ and $\frac{3}{4} \times 400 = 300$.

Similarly, in dihybrid inheritance, a 9 : 3 : 3 : 1 ratio means each one of the offspring has 9 : 3 : 3 : 1 chance of having any of the four phenotypes. If there were a population of 320, to find how many in each class in the ratio, find how many in the class of 1: 9 + 3 + 3 + 1 = 16

$$320 \div 16 = 20$$

So the class of 1 has 20. The classes of 3 each have 3 × 20 = 60 and the class of 9 has 9 × 20 = 180.

Alternatively, using fractions, 9 : 3 : 3 : 1 is the same as $\frac{9}{16} : \frac{3}{16} : \frac{3}{16} : \frac{1}{16}$. Then $\frac{1}{16} \times 320 = 20$, $\frac{3}{16} \times 320 = 60$ and $\frac{9}{16} \times 320 = 180$.

quickfire 5.13

a) In a cross between a black mouse, Bb, and a white mouse, bb, how many black and how many white baby mice would you expect in a litter of 8?

b) In tomato plants, P is the allele for purple stems and H is the allele for hairy stems. In a cross between two plants, PpHh, with purple, hairy stems, how many offspring would you expect to have purple, hairy stems out of a total of 480?

5.4 Theoretical ratios and real life

The ratios described above are produced in ideal situations. They assume all gametes are equally available and that all are equally able to fuse. They also assume that all zygotes have an equal probability of surviving to the point where they are counted. Chance plays a role in which gametes fuse together, and the assumptions may not really be true. The result is that the proportions observed rarely match the ideal ratio. This is a statistical phenomenon and the larger the sample, the more closely the observed ratio is to the ideal.

You are expected to look at reported numbers of offspring of various phenotypes and decide which ratio they most closely match. In a cross between black and white guinea pigs, all the F_1 are black. In the F_2, if the offspring all mated and

A black guinea pig

produced 31 black pups and 9 white, you might guess that the ideal ratio was 3 : 1 and assume this is standard monohybrid inheritance:

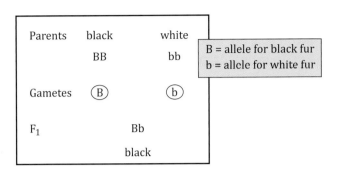

A white guinea pig

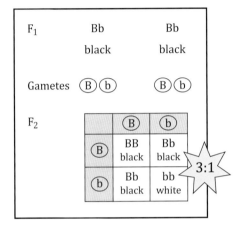

quickfire ▸▸▸ **5.14**

What is the ideal ratio of a population of 485 tall wheat plants and 515 dwarf?

quickfire ▸▸▸ **5.15**

What is the ideal ratio of a population of people with 730 whose blood is Rhesus positive and 270 whose blood is Rhesus negative?

Similarly, if a tomato plant with a hairy stem and dissected leaf margins were crossed with one with a hairless stem and smooth leaf margins you might find these proportions:

98 hairy, dissected : 102 hairy, smooth : 104 hairless, dissected : 97 hairless, smooth.

You would then guess that the ideal ratio is 1 : 1 : 1 : 1 and that the cross is a dihybrid test cross:

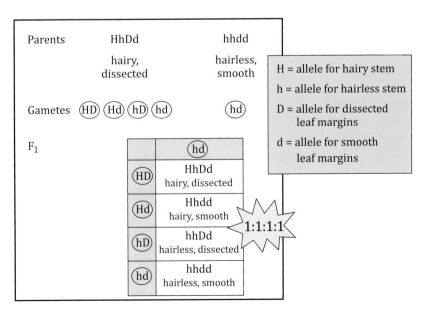

A tomato plant

quickfire ▸▸▸ **5.16**

What are the genotypes of the parents of a population of moss plants of which 102 have broad, blunt leaves, 98 have broad, pointed leaves, 92 have narrow, blunt leaves and 108 have narrow pointed leaves?

Dexter

Kerry

quickfire 5.17

Cattle breeders who want to breed Dexter calves prefer to cross Kerries with Dexters, rather than inter-breed the Dexters. Why?

5.5 Non-Mendelian ratios

Sometimes the results of a cross do not seem to correlate with any of the ideal Mendelian ratios. There may be many reasons for this. The assumptions described on page 100 may not be true in a particular case; it may be that gametes have unequal survival or unequal ability to fuse in fertilisation. There may be other reasons, as described below.

5.5.1 Lethal recessives

A common cause of non-Mendelian ratios is related to zygote viability. In these cases, if a zygote is homozygous for a particular recessive allele, it does not survive. Cattle provide an example:

Kerry cattle are homozygous for the D gene, DD, and have legs the normal length. Dexter cattle have short legs and are heterozygous at this locus, so they have the genotype Dd. The genotype dd generates the 'bulldog' phenotype. If you cross two Dexter cattle, the expected phenotypic ratio in the offspring would be 1 Kerry : 2 Dexter : 1 bulldog:

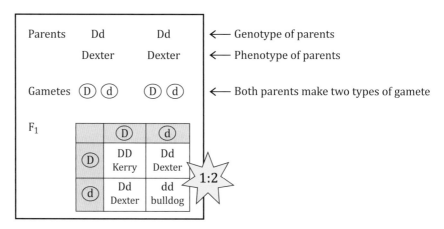

But dd is a lethal combination and so a cow is unlikely to give birth to a live calf with the bulldog phenotype. The ratio of surviving offspring is therefore 1 Kerry : 2 Dexter. This is not a Mendelian ratio, but is explained by applying Mendelian genetics when a lethal combination of recessive alleles occurs.

5.5.2 Epistasis

The examples of genetics so far have described characteristics affected by one gene. But there are situations where more than one gene can affect a single characteristic, and then, non-standard ratios are seen. An example is the genetic control of coat colour in many mammals. Mice have been studied most because they are easily kept and have a short generation time, but it has also been shown in rabbits, horses and others. A similar situation occurs in the pigmentation of some plants, such as in the colour of maize seeds.

Epistasis is also called gene suppression, because one gene, the epistatic or inhibiting gene, controls whether or not the other, the hypostatic gene, is expressed.

Mice can be coloured or albino. The epistatic gene controlling this is designated C. So a CC or Cc mouse is coloured and a cc mouse is albino. If the dominant allele, C, is present, then the hypostatic gene, B, for colour can be expressed. In the presence of C, BB and Bb mice are black and bb mice are brown.

The phenotypes produced by these combinations of allele are shown in the table:

		Genotype at gene C		
		CC	Cc	cc
Genotype at gene B	BB	black	black	albino
	Bb	black	black	albino
	bb	brown	brown	albino

In a dihybrid cross, where two black mice heterozygous at both genes, CcBb, are crossed, the Mendelian ratio 9 : 3 : 3 : 1 is not seen as there are only three phenotypes, black, brown or albino, not four, as usual when two genes are analysed. The grid below shows this:

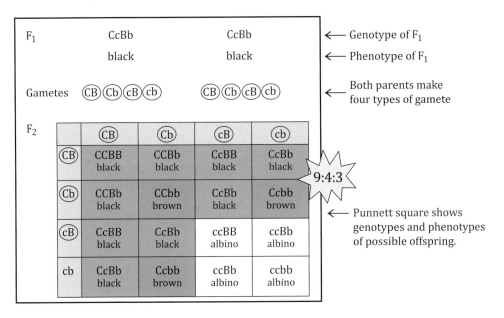

The ratio is 9 black : 3 brown : 4 albino. It is not one of the Mendelian, ideal ratios, but it has been explained using Mendelian genetics.

When a dihybrid test cross is made, one of the parents is recessive at both genes, in this case ccbb. A cross with a double heterozygote, CcBb, would give, in normal circumstances, a 1 : 1 : 1 : 1 ratio of the four possible phenotypes. In this cross, only three phenotypes are available so a non-Mendelian ratio is seen:

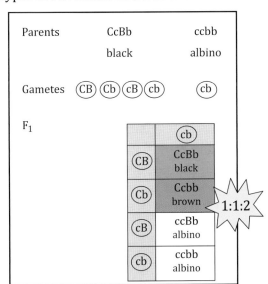

Mice

>> **Pointer**

When you are writing allele symbols and the capital letter has the same shape as the small letter, e.g. C and c, make sure the capital letter is written much bigger than the small letter so that you do not get confused.

quickfire >> 5.18

In maize, the dominant allele A gives purple grains and the recessive, a, gives red grains. The A/a gene is only expressed in the presence of the dominant allele E of the epistatic gene E/e. White grains are produced by the ee genotype. What genotypes for these two genes produce white grains?

Red sweet pea flower

5.5.3 Linkage

When Mendel analysed the behaviour of two genes at the same time in his dihybrid crosses, he made lucky choices. The genes he chose were on different chromosomes and behaved independently of each other. That was why the parents formed equal numbers of the different gamete types. The equal probability with which they all fused gave the standard $9:3:3:1$ ratio.

At the beginning of the 20th century, genetics still used sweet peas as a model organism, as Mendel had done. In a famous experiment, Punnett crossed sweet peas with purple flowers and long pollen grains with plants that had red flowers and short pollen grains. If the genes behaved in a Mendelian fashion, a $9:3:3:1$ ratio in the F_2 offspring would have been expected. But in this cross, most of the F_2 had either purple flowers and long pollen grains or red flowers and short pollen grains. These phenotypes were described as parental, because that is what was seen in the parents. But sometimes, offspring had purple flowers and short pollen grains or red flowers and long pollen grains. These were called recombinant phenotypes, because the characteristics in the parents had recombined to give new phenotypic combinations. Here is a diagram to show the family tree, or pedigree:

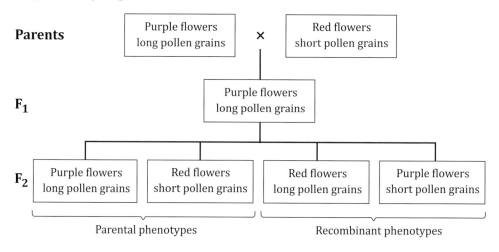

Purple sweet pea flower

Here are the alleles for the two genes:

P = allele for purple flowers; p = allele for red flowers

L = allele for long pollen grains; l = allele for short pollen grains

The non-Mendelian ratio can be explained by the two genes being on the same chromosome. They are physically linked and likely to be inherited together. As a result, the different gamete types were not formed in equal numbers, so the F_2 ratio was not $9:3:3:1$. When two plants, PPLL and ppll cross, their F_1 will be PpLl. If the P and L genes are close together on the same chromosome, most of the gametes will be PL and pl so most of the F_2 will be PPLL, ppll or PpLl.

There will be a small amount of crossing over during gamete formation, so there will be a small number of Pl and pL gametes. This means that all the other possible genotypes will occur but in much smaller numbers than if all gamete types were made with equal frequency.

The cross is shown on page 105.

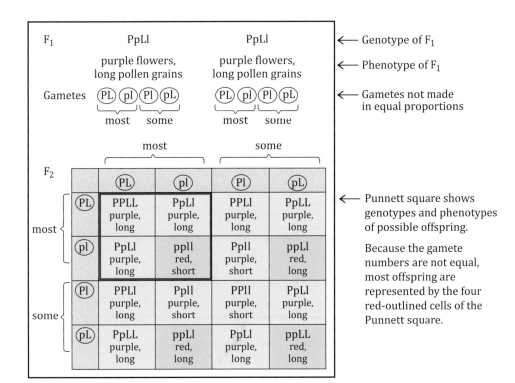

← Genotype of F$_1$

← Phenotype of F$_1$

← Gametes not made in equal proportions

← Punnett square shows genotypes and phenotypes of possible offspring.

Because the gamete numbers are not equal, most offspring are represented by the four red-outlined cells of the Punnett square.

>> *Pointer*
When a non-standard ratio of four phenotypes is seen, the reason is often genetic linkage.

Recombination frequency

If two genes are unlinked, then the F$_2$ phenotypic ratio is 9 : 3 : 3 : 1. If two genes are completely linked so that no cross-overs occur between them, they behave as one gene and the F$_2$ phenotypic ratio will be as for a monohybrid cross , i.e. 3 : 1, with none of the recombinant phenotypes. If there are some cross-overs then the proportion of recombinants is an indication of the frequency of cross-over, and, therefore, of the distance between the genes. The further apart they are, the more cross-overs are possible, the more F$_2$ offspring with recombinant phenotypes will be produced.

The recombination frequency (RF) between two genes can be calculated:

$$RF = \frac{\text{number of recombinants}}{\text{total number of progeny}} \times 100\%$$

A well-known example concerns two maize genes, one for shape (Sh = shrunken, sh = plump) and one for colour (A = coloured, a = colourless). A cross between parents who are either homozygous dominant (AA ShSh) or homozygous recessive (aa shsh) at both genes produces offspring that are heterozygous at both genes(Aa Sssh). A cross between two of these produced the following:

Phenotype	Number	Status
coloured shrunken	5020	parental
colourless plump	4960	parental
coloured plump	12	recombinant
colourless shrunken	8	recombinant

Using the equation, $RF = \dfrac{\text{number of recombinants}}{\text{total number of progeny}} \times 100\%$

$$= \frac{12 + 8}{5020 + 4960 + 12 + 8} = \frac{20}{10\,000} \times 100 = 0.2\%$$

quickfire>> 5.19

In a cross between tomato plants homozygous for the genes producing either green, hairless stems or purple, hairy stems, all the F$_1$ had purple, hairy stems. The F$_2$ from crossing these F$_1$ produced the following numbers of offspring: 293 with purple, hairy stems, 15 with purple, hairless stems, 12 with green, hairy stems and 98 with green, hairless stems. What can you infer from these numbers?

Map units

A map of the relative positions of genes can be made using their recombination frequencies as a measure of the distance between them. A map unit (1 mu) is defined as the distance between pairs of genes that have RF = 1%. So if two genes have RF = 0.2%, they are 0.2 mu apart. A map unit is sometimes given the unit cM (centiMorgan).

Genetic mapping

If we have genes A, B and C and it is shown that A and B are 10 mu apart they can be placed on a putative chromosome:

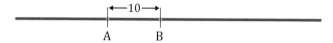

If B and C are 20 mu apart, C could be in either of two positions:

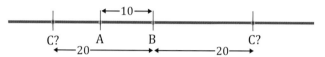

The position of C can be ascertained by finding the distance between A and C. It could be to the left of A, as drawn, in which case the distance A–C = 20 – 10 = 10 mu. If C is to the right of A, the distance A–C = 20 + 10 = 30 mu:

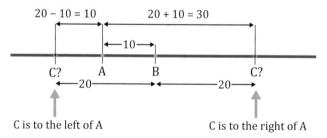

Double cross-overs and mapping

There could be two cross-over events between two linked genes, as shown below. The resulting chromatids would still have the allele combinations AB or ab. Looking at the progeny for recombinants would not tell you a cross-over had occurred, unless you brought in a third gene.

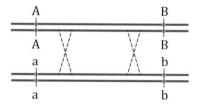

If we consider three linked genes, A, B and C, there could be cross-overs between A and B and between B and C, as shown below.

With two cross-overs, the possible allele combinations would be AbC and aBc.

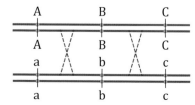

In a real cell, there would also be chromatids formed from one cross-over (ABc, abC, aBC and Abc) and from no crossovers (ABC and abc). Double cross-overs would happen much less often than a single or no cross-overs so the chromatids AbC and aBc would be fewest. These chromatids would tell you that gene B is in the middle. It would then be possible to map the genes by working out the recombination frequencies between A and B and between B and C.

5.5.4 Sex linkage

Genes on the sex chromosomes may code for characteristics that are expressed preferentially in one sex or the other. Well-known examples in humans are genes associated with some types of haemophilia, Duchenne muscular dystrophy and with colour-blindness. In these cases, the recessive allele on the X chromosome is expressed in males as there is no homologous region on the Y chromosome to carry a dominant allele. So these conditions are expressed predominantly in males. Females with two recessive alleles would also have the condition.

A tortoiseshell cat

There is a gene X^o on the X chromosome of cats which has an effect on coat colour. The X^o allele produces orange fur and the X^B allele produces black fur. Thus, a male could be orange or black. Females have one of three genotypes $X^o X^o$, $X^o X^B$ or $X^B X^B$. The $X^o X^B$ genotype produces both colours and the cats are described as tortoiseshell. The result of a cross between a tortoiseshell female and a black male produces a non-Mendelian ratio:

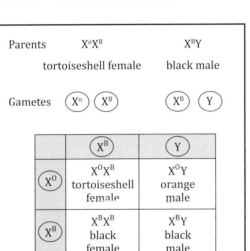

Parents	$X^o X^B$	$X^B Y$
	tortoiseshell female	black male
Gametes	X^o X^B	X^B Y

	X^B	Y
X^o	$X^o X^B$ tortoiseshell female	$X^o Y$ orange male
X^B	$X^B X^B$ black female	$X^B Y$ black male

As the Punnett square shows, the ratio overall is 1 tortoiseshell : 1 orange : 2 black. This is reminiscent of the 1 : 2 : 1 ratio signifying incomplete dominance but it is not incomplete dominance, because the heterozygote is in the class of 1, not the class of 2.

It is more useful to look at the males and females separately. The females show a ratio of 1 tortoiseshell : 1 black and the males show 1 orange : 1 black.

» Pointer
Sex linkage produces non-Mendelian ratios because in males, only one allele contributes to the phenotype, not two.

Summary

This chapter has shown how the Mendelian ratios are derived. It has explained some deviations from those ratios and introduced the concept of genetic mapping. A cross is vanishingly unlikely to produce progeny in the numbers predicted, and whether a particular set of counts represents a particular cross can be inferred by a statistical analysis of the results. This will be described in a later chapter.

Test yourself 5

❶ The colour of chocolate mice is controlled by the gene C. The recessive allele, c, is associated with white fur colour. Two chocolate mice mate and produce a litter containing 6 chocolate mice and 2 white mice.

    a) What ratio do the mice in the new litter represent?

    b) What are the genotypes of the parents likely to be?

    c) Draw a Punnett square to show how these parents could produce such a litter.

    d) If these mice had a further litter of four, how many would you expect to be white?

❷ If a plant homozygous for green leaves and a plant homozygous for yellow leaves are crossed, all their offspring all have green leaves. What is the outcome if one of the offspring is crossed with a homozygous yellow plant?

❸ Some black cats have small white patches because of the action of the gene S. Other cats have large white patches or no patches. The S gene, which controls the patches, has two co-dominant alleles, S^1 and S^2.

A black cat without patches has the genotype S^1S^1. It was mated with a cat with small white patches. Some of the kittens had small white patches and some were entirely black.

Draw a genetic diagram to show how this could have happened.

❹ Pure-breeding thorn-apples were crossed. One had mauve flowers and smooth fruit and the other had white flowers and prickly fruit. All their F_1 had mauve flowers and prickly fruit. These were self-fertilised and produced the following:

    mauve flowers, prickly fruit   925

    mauve flowers, smooth fruit 290

    white flowers, prickly fruit   294

    white flowers, smooth fruit   105

Draw genetic diagrams to explain these observations.

❺ Two characteristics are separately controlled by the two unlinked genes R and S.

    a) Draw a Punnett square to show the outcome of the cross RrSs × rrss

    b) If the genes were linked and there were a cross-over event between them, how might the pattern of inheritance be different?

❻ Genes for eye colour and wing length in *Drosophila* are linked. The alleles can be shown as R = red eyes; r = purple eyes and N = normal wings; n = vestigial wings. In a cross between two individuals, RrNn × rrnn, the following were produced:

    Rrnn   165

    RrNn   191

    rrNn   23

    Rrnn   21

Find the recombination frequency between the genes.

7 Guinea pigs can have rough fur, produced by the dominant allele R, or smooth fur, produced by the recessive allele r. The allele for black fur (B) is dominant to the allele for white fur (b).

A cross between guinea pigs with genotypes BBRR and bbrr produces an F_1 with genotype BbRr. When these guinea pigs are interbred, the F_2 are produced as shown below:

black, rough fur   58

black, smooth fur   6

white, rough fur   4

white smooth fur   20

a)   What ratio of phenotypes would have been expected from this cross?

b)   Why has the expected ratio not been produced?

c)   What can you infer about the chromosomal positions of the two genes described here?

8 Some fruit flies have red eyes because they have an allele on the X chromosome, X^R. This is dominant to the allele for white eyes, X^r. Males have only one allele for this gene as they have only one X chromosome. Their Y chromosome is small and does not carry the gene that codes for eye colour.

Show a cross between a white-eyed female and a red-eyed male, giving the gametes produced by each and the genotypes and phenotypes of their offspring.

9 Draw a map for the four genes K, L, M and N given that the RF values for the pairs of genes are: K–M = 15%, K–N = 25%, M–L = 18% and L–N = 8%.

Chapter 6

The Hardy–Weinberg Equilibrium

≫ Pointer

The allele frequency, written as a decimal, shows how often the allele occurs, out of all the alleles for that gene, in the population.

quickƒire ≫ 6.1

For a particular gene in the gene pool of a population, 55% of the alleles were dominant.
What is the allele frequency?

quickƒire ≫ 6.2

The frequency of a dominant allele M is 0.6. What is the frequency of the recessive allele, m, of that gene?

≫ Pointer

Dominant alleles do not necessarily have a high frequency and recessive characteristics do not necessarily have a low frequency.

quickƒire ≫ 6.3

If the allele frequency for a dominant allele H is 0.6, what is the frequency of the genotype AA?

quickƒire ≫ 6.4

If the allele frequency for a dominant allele D is 0.8, what is the frequency of the genotype dd?

quickƒire ≫ 6.5

If the frequency of a dominant allele K is 0.61 and the frequency of the recessive allele is 0.39, what is the frequency of carriers in the population?

The Hardy–Weinberg equilibrium describes the frequency of alleles in a population. They are usually defined as the dominant allele A and the recessive allele a. 'Equilibrium' refers to the fact that their frequencies stay the same down the generations, under particular conditions. When certain assumptions are made, the frequencies of the alleles can be calculated using the Hardy–Weinberg equation, the only use of a quadratic equation that you are likely to meet in A Level Biology.

Whenever this equilibrium is discussed, the symbol p is used for the frequency of a dominant allele, which, by convention, has the symbol A. The symbol q is used for the frequency of the recessive allele with the symbol a. That convention will be followed here. The logic is the same for plant and animal populations, but only animals will be mentioned here, for convenience.

The per cent frequency of an allele is the percentage of that allele out of all the alleles for that gene in the population. The frequency of an allele shows how often it occurs out of all the alleles for that gene in the population and it is expressed as a decimal.

If there are two alleles for the gene we are considering, one dominant and one recessive, then genotypes of all the individuals in the population are AA, aa or Aa. As there are only A and a alleles, the frequencies of the two add up to 1, as that is all there is. In other words: $p + q = 1$. There is no relationship between dominance and allele frequency, so a dominant allele may be rare and a recessive allele common. Dominant characteristics do not necessarily spread themselves throughout a population.

AA individuals occur when an egg carrying the allele A and a sperm carrying the allele A fuse. The frequency of an allele in the population as a whole is the same as its frequency in a sample of the population. So the probability of an egg having A is p and the probability of a sperm having A is also p. Consequently, the probability of AA individuals arising will be the frequency of A multiplied by the frequency of A. This can be written as $f(AA) = p^2$, where f means 'the frequency of'. Similarly, the probability of an egg carrying the a allele is q and the probability of a sperm having the a allele is q so the probability of aa individuals arising will be the frequency of a multiplied by the frequency of a. This can be written as $f(aa) = q^2$.

A heterozygous individual could have A from the egg and a from the sperm. The frequency of such an individual would be $p \times q$. But they could have been formed the other way round, with a from the egg and A from the sperm. The frequency of this second type is also $p \times q$. So the total frequency of all Aa individuals is $(p \times q) + (p \times q) = 2pq$.

These last two paragraphs have described the whole population, so if we add the frequencies of AA, aa and Aa, they must equal 1. In other words, $p^2 + 2pq + q^2 = 1$ or $(p + q)^2 = 1$. This is the Hardy–Weinberg equation. After a series of random matings, the ratio of the three genotypes will still be $p^2 : 2pq : q^2$. The values of p and q have not changed, so in every generation the ratio holds. The frequencies of the dominant and recessive allele are constant, and this is the equilibrium described by the Hardy–Weinberg equation.

Real populations may actually have variation in allele frequencies in successive generations but in an ideal population, there will be none and the allele frequencies will stay the same. To be 'ideal' a population must be large and with no immigration or emigration and no mutation. Individuals mate at random. All gametes are assumed to be equally fertile and all genotypes have equal chance of survival.

Using the equation

Knowing the frequency of certain inherited conditions in the population allows you to calculate the frequency of the allele concerned. Conversely, knowing the frequency of alleles can allow you to predict the frequency of conditions. This is an aspect of medical genetics and if you have a future as a genetics counsellor, such concepts will be important to you.

Here is an example where we are told the frequency of a condition associated with a homozygous recessive allele. The frequency of the recessive allele (q) is found from this and then the frequency of carriers ($2pq$).

> One person in 10 000 is born with the inherited disease phenylketonuria (PKU). It occurs when a child inherits two recessive alleles. Find (a) the frequency of the recessive allele and (b) the proportion of carriers in the population.
>
> (a) PKU is associated with a recessive allele. Let the allele symbol be a.
>
>   Then $f(a) = q$. (This is what we are trying to find.)
>
>   Individuals with PKU have the genotype aa.
>
>   The frequency of affected individuals $= q \times q = q^2$.
>
>   1 person in 10 000 has the condition
>
> $\therefore q^2 = \dfrac{1}{10\,000} = 0.0001$
>
> $\therefore q = \sqrt{0.0001} = 0.01$
>
> (b) The proportion of carriers $= 2pq$. (This is what we are trying to find.)
>
> $q = 0.01$
>
> $p + q = 1$
>
> $\therefore p = 1 - q = 1 - 0.01 = 0.99$
>
> $\therefore 2pq = 2 \times 0.99 \times 0.01 = 0.198$

In the calculation above, the question asks for the proportion of carriers. This is the same as the frequency. A question might ask for a percentage of carriers. This is another of those situations where you have to read the question very carefully to ensure you know exactly what is asked. If the percentage had been needed, all you do is multiply the frequency or proportion by 100:

Frequency of carriers $= 0.198$

$\therefore$% carriers $= 0.198 \times 100 = 19.8\%$

» Pointer

p and q are allele frequencies. p^2, $2pq$ and q^2 are genotype frequencies.

quickfire » 6.6

The frequency of a dominant allele R is 0.3. What is the frequency of heterozygotes, Rr?

» Pointer

% = frequency × 100.

Test yourself 6

1 PTC is phenyl thiocarbamate. Some people can taste it and some people cannot. The ability to taste is conferred by the dominant allele of a single gene. In a population of people, 0.2% cannot taste PTC and are called 'non-tasters'.

a) What is the frequency of non-tasters in the population?

b) What is the frequency of the recessive, non-taster allele?

c) What is the frequency of the dominant, taster allele?

d) What is the frequency of heterozygotes?

e) What is the frequency of tasters in the population?

2 About 1 person in 200 has type I diabetes and fails to secrete adequate insulin. This condition can be controlled by a single gene which produces diabetes when homozygous. What is the frequency of the carrier genotype?

3 The MN blood group system comprises two co-dominant alleles. They produce the genotypes MM, NN and MN and their respective blood groups M, N and MN. In a population 58% of people were blood group M and 36% blood group MN. Calculate the frequency of alleles M and N in the population.

4 People who can roll their tongue have the dominant allele R. People who have two recessive alleles, the genotype rr, cannot roll their tongue. It was estimated that 64% of people could roll their tongue. Calculate the percentage of genotypes RR, Rr and rr in the population.

Chapter 7

Statistics

As you may have read in the Introduction, this is not a Mathematics text book. Nor is it a Statistics text book, but this chapter will allow you to face the statistics of A Level Biology with confidence. It has been said that you only really need statistics when the answer isn't obvious. That is not the view of your examiners at A Level so even if the data do strongly suggest a conclusion, a statistical analysis will support your prejudices; or perhaps not.

The need for statistical analysis of data arises from the inherent variation between living organisms. They do not necessarily behave in the same way, despite all attempts to control experimental variables. In collecting data, it is important to remember that chance and probability make a contribution. Identical readings are therefore treated with some suspicion, a point often not appreciated by students in practical examinations. Statistical tests allow a judgment on whether a set of readings is likely to be due to a biological phenomenon or merely due to chance.

7.1 Data

Data are fundamental. Raw data are the actual readings or the counts that you make, such as the distance moved by an air bubble in a potometer in 5 minutes. The data can be processed to give you some other parameter which may be more meaningful, such as multiplying the distance moved by the bubble by πr^2 to calculate the volume of water taken up into a shoot. However, the statistical tests that you are likely to use were designed to use raw data only.

7.2 Sampling

An experiment assessing a population will not be able to use data applying to every member of the population. The population must be sampled but the sample must reflect the whole population.

a) If sampling is random, as in many fieldwork experiments, every member of the population, or every co-ordinate examined, has an equal chance of being selected, such as when comparing per cent cover of a species in different habitats.

b) Stratified sampling divides the population into sections, each of which is sampled, such as when testing the effect of nitrate fertiliser on cereal growth. In this case, each species should be examined separately.

c) Block sampling is used if an area is known to have significant differences, such as light intensity on some areas of a field. Then the field is subdivided so that only those areas that are similar will be sampled.

> **Pointer**
> Statistical tests are performed on raw not manipulated data.

quickfire 7.1

Raw or processed data?

(i) Mass of potato chip after submersion in sucrose solution for 24 h.

(ii) Percentage increase in mass of potato chip after submersion in sucrose solution for 24 h.

(iii) Number of bubbles coming from submerged water weed in a 5-minute observation period.

(iv) Rate of bubble production (bubbles/minute) in submerged water weed in a 5-minute observation period.

(v) Number of beetles in a pit-fall trap.

(vi) Lincoln index.

d) If, on the other hand, care is taken that the sample reflects certain aspects of the whole population, sampling is described as proportional, such as when assessing susceptibility to blood-borne disease. In this case, it may be useful to ensure that the sample has the same proportions of different blood groups as the whole population.

7.3 Probability

»Pointer

Probability can be predicted or can be calculated as:

$$\frac{\text{number of successes}}{\text{number of trials}}.$$

quickfire» 7.2

Estimate the probability of:

(i) A cross between two black guinea pigs, BB and Bb, having only black pups in their litter, where B is the allele for black fur and b is the allele for white fur.

(ii) A cross between two right-handed people, Rr and RR, having a left-handed child, where R is the allele for right hand dominant and r is the allele for left hand dominant.

(iii) A freckled child of a parent with freckles, Ff, and a parent who does not make freckles, ff, where F is the allele for producing freckles and f is the allele for not producing them.

(iv) A seed producing a white-flowered plant from a cross between two pea plants, both with mauve flowers (Aa), where A is the allele for producing the pigment anthocyanin, and a is the allele for not producing the pigment.

»Pointer

Probability is expressed as a decimal between 0 and 1.

In Biology, we use statistical tests to say how likely it is that our results have a biological explanation or whether they are due to chance. In other words, we are asking if it is probable that our data have biological meaning. The mathematical field of probability calculates the likelihood of events happening and that likelihood is given a numerical value. The word 'chance' is not a mathematical word but is used in ordinary language to mean the same thing in a less precise way.

There are two ways of estimating a probability. In some situations, based on biological theory, we can predict the probability of an event, such as the outcome of a cross between a tall plant, with genotype Tt, and a dwarf plant, tt. In other cases, we have to measure it, such as estimating the probability of one seed from a batch germinating.

In this case, the probability $p = \dfrac{\text{number of seeds germinated}}{\text{total number of seeds}}$.

In general, if you have to calculate a probability, you can use the equation

$$p = \frac{\text{number of successes}}{\text{number of trials}}.$$

Probability is expressed as a decimal. Apart from the field of quantum mechanics, which does not trouble us here, we can say that if an event is impossible, $p = 0$ and if it is inevitable, $p = 1$. Anything in between is quoted as a decimal between 0 and 1.

7.4 Averages

When people use the word 'average', they are usually referring to the mean. In scientific writing, it is best to be absolutely unambiguous, so if you mean the mean, say so, because the mode and the median could also be referred to as averages.

7.4.1 Arithmetic mean

The mean used in Biology is the arithmetic mean. It is the sum of all values divided by the number of values. It is often given the symbol $\bar{x}$. The calculation of the mean of three values, x_1, x_2 and x_3 can be written as

$$\bar{x} = \frac{x_1 + x_2 + x_3}{3} \text{ or as } \bar{x} = \frac{\Sigma x}{n}, \text{ where } \Sigma x \text{ means 'the sum of all values of } x'.$$

The problem with using a mean to describe a set of data is that it includes all values, even the anomalous and extreme values. If there were one value which was much higher than all the others, this would make the mean artificially, and perhaps, incorrectly, high. If there were one value which was much lower than all the others, this would make the mean artificially and perhaps, incorrectly, low. This is why we always take many readings in a biology experiment. Chance always plays a part and anomalous results do occur. If an arithmetic mean is used, the effect of the anomaly is minimised and so the mean is more reliable than an individual reading.

7.4.2 Median

The median is the middle value. There is an equal number of values above and below it. When there is an odd number of values, it is straightforward to work out which is the middle one. The values can be put in order and then you see which value is halfway down. For example here is a list of 7 values for the length of nettle leaves: 11 mm, 37 mm, 58 mm, 21 mm, 45 mm, 52 mm, 47 mm. They can be put in order: 11, 21, 37, 45, 47, 52, 58. There are 7 values so the median will be the fourth, 45 mm, and there are three values above and three below.

But if there is an even number of values, there will be two in the middle. The median is then the mean of these two. Here is a list of 8 lengths of limpet shells: 24 mm, 29 mm, 18 mm, 9 mm, 13 mm, 39 mm, 22 mm, 8 mm. The values can be put in order: 8, 9, 13, 18, 22, 24, 29, 39. The middle two are the fourth and fifth, 18 and 22. The median of the whole set of values is the mean of these two:

$$\frac{18 + 22}{2} = \frac{40}{2} = 20 \text{ mm.}$$

The median is useful because it is not skewed by extreme values. The median of 2, 4, 6, 8, 100 is the same as the median of 2, 4, 6, 8, 10.

7.4.3 Mode

The mode is the value which occurs most often in a data set. Here is a set of counts of the number of spots on a group of seven ladybirds: 2, 4, 8, 8, 10, 4, 8. Observation shows that 8 spots is the most frequent and so the mode is 8. If the data set is too large to assess by eye, a tally chart will show which is the largest class:

Number of spots	Number of ladybirds
2	1
4	2
8	3
10	1

Mode = 8 ⟶

7.5 Distributions

7.5.1 Normal distribution

In Chapter 4, continuous variation was discussed and it was explained how this can be shown graphically as a normal distribution curve. If a biological characteristic is normally distributed, such as human height, the curve is symmetrical and the mean, median and mode are identical. The extremes of the curve are called the tails and are indicated on the diagram at the top of page 116.

>> *Pointer*
Mean = $\frac{\text{sum of all values}}{\text{number of values}}$.

>> *Pointer*
Median = middle value.

>> *Pointer*
Mode = commonest value.

quickfire >> 7.3
Here are five values for the swimming speed in mm / second of crab larvae: 1.0, 2.1, 3.6, 5.5, 7.8.
(i) Calculate the mean swimming speed.
(ii) What is the median swimming speed?
(iii) Here are the areas in mm² of leaves of ground ivy. What area range is the mode?

Leaf area /mm²	Number of leaves
0–19	3
20–39	9
40–59	12
60–79	17
80–99	14

» Pointer

In a normal distribution, the mean, median and mode are identical.

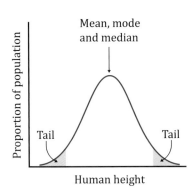

7.5.2 Negative skew

Some characteristics may show more extremely small values than extremely large and so the curve that describes them is not symmetrical. The left tail, representing small values, is much longer than the right tail. The curve is negatively skewed. An example is the length of cod in the North Sea and may be related to smaller fish escaping through the mesh in the fishing nets, as shown below.

» Pointer

A negative skew has more small values than large.

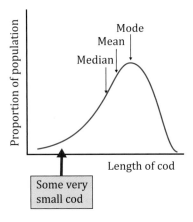

The mode is easily identified as the length most represented. The median is lower than the mode because there are more low values than high values. The mean on this curve is between the mode and the median. But if the tail is extremely long, the mean could be even lower than the median, as it would be influenced by a lot of very short fish and only a few very long ones.

7.5.3 Positive skew

Some characteristics may show more extremely large values than extremely small ones and so the curve that describes them is not symmetrical. The right tail, representing large values, is much longer than the left tail. The curve is positively skewed. An example is the number of puppies in a litter. Most Yorkshire terriers produce up to three puppies in their first litter, but may produce more, occasionally even six.

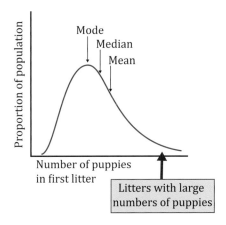

» Pointer
A positive skew has more large values than small.

The mode is easily identified as the number of puppies in a litter that occurs most often. The median is higher because there are more very high values. The mean even higher, as it is influenced by the high number of puppies in some litters.

7.5.4 Bimodal distribution

Bimodal means having two modes. It suggests that there is an advantage to having a small value for a characteristic, or a large value, but not an intermediate one. An example could relate to the lengths of fish. Long fish can produce more propulsion than short ones and can quickly swim away from predators, avoiding being eaten. Short fish may be more easily camouflaged in weed and also escape predation. Thus the intermediate length fish are selected against and the population shows two distinct lengths. There is variation in the length of the long fish and of the short fish so this population may be thought of as a combination of two normally distributed sub-populations.

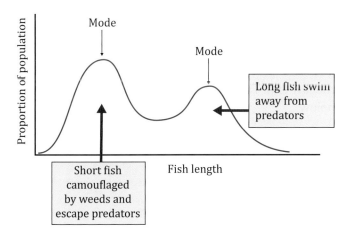

» Pointer
A bimodal distribution has two modes.

7.6 Variability

Experimental readings have to be repeated and a mean is calculated. This is because there is inherent variation and a mean is more reliable than a single reading. But a mean gives only one piece of information about a set of readings and there are additional aspects of the readings that may be described to give more meaning.

quickfire >> 7.4

Calculate the means and state the median values for the following data:

(i) On five successive days, the numbers of rabbits seen in a 1 hectare field were 46, 37, 49, 23, 19.

(ii) The serum cholesterol in five healthy people was measured as 0.69, 1.48, 0.97, 0.54 and 1.42 mmol dm⁻³.

(iii) The numbers of mayfly nymphs caught in five kick samples was 23, 17, 9, 11, 20.

7.6.1 Range

Range refers to the difference between the smallest and largest readings in a data set. It uses only a small part of the data and says nothing about the distribution within the range. You might use range bars when you plot data points for a line graph or on the bars of a histogram or bar chart to indicate the range contributing to the mean.

7.6.2 Standard deviation

Standard deviation takes into account all the readings and shows their spread around the mean. The middle of the slope either side of the mean marks the size of the standard deviation. Two populations are shown as normal distributions in the diagrams below, with their standard deviations indicated below the *x* axes:

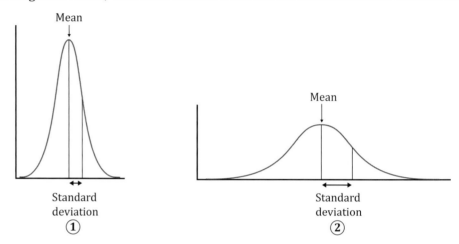

The curve on the left appears narrower than the curve on the right. This means that the readings for the curve in the left are all closer to the mean than for the curve on the right. This is one way of explaining what is meant by saying that standard deviation ① is much smaller than standard deviation ②. In a normal distribution, more readings are closer to the mean than further away from it. The graph below shows that 68% of all values are within one standard deviation of the mean and 95% are within 2. This means that 5% of values are on the tails, 2.5% at the high end and 2.5% at the low end. The values are represented by the area under the graph, as shown below:

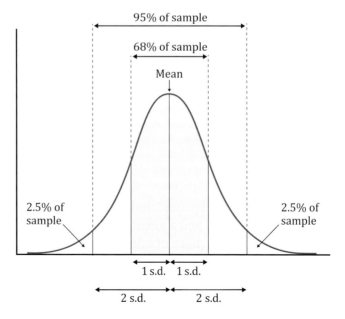

The area under the curve represents the whole population. This means that specified areas can be thought of as the probability that a single sample will fall within that range, e.g. the area within one standard deviation of the mean is 68% so the probability of a single sample being within one standard deviation of the mean is 0.68.

» Pointer
Standard deviation describes the spread of readings around their mean.

7.6.3 Variance

The variance is used when you test the difference between means using a t test. This is how to understand it. If you have a set of readings, some are above the mean and some are below. Those above the mean have a positive deviation from the mean and those below have a negative deviation. If you add up how much they deviate from the mean, the positives cancel out the negatives and the answer is zero. Zero is not much use to statisticians. So they square the deviations, because when you square negative numbers, the answer is positive. When you add up all the squares of the deviations you get a positive answer, which is much more use. Divide this by the number of readings and you have the variance. It is the square of the standard deviation.

Here is how to do it:

If the mean is $\bar{x}$ then deviation from the mean is $(\bar{x} - x)$.

The deviation squared is written $(\bar{x} - x)^2$.

The sum of the deviations from the mean is written as $\Sigma(\bar{x} - x)^2$.

The number of readings is n.

The variance has the symbol s^2, so $s^2 = \dfrac{\Sigma(\bar{x} - x)^2}{n}$.

Another equation to calculate variance is sometimes used. It is written

$s^2 = \dfrac{\Sigma x^2}{n} \; \bar{x}^2$. Both these formulations appear in this book.

Calculating standard deviation

» Pointer
Variance = standard deviation^2.

The standard deviation is the square root of the variance. It refers to the distribution of individual values. You might use standard deviation bars around the data points on a line graph or on a histogram or bar chart to indicate the spread around the mean of the sample. On page 120 you can see how to calculate variance and standard deviation from experimental data.

7.6.4 Standard error

Standard error is sometimes referred to as 'standard error of the mean' because unlike standard deviation, which refers to uncertainty of individual readings, standard error refers to uncertainty of the estimate of the mean. It is calculated as

$$\sigma_{\bar{x}} = \frac{s}{\sqrt{n}}$$

where $\sigma_{\bar{x}}$ = standard error of mean, s = standard deviation and n = number of values.

It should be quoted with a + sign, as should standard deviation. Both are used, for example, on bar charts. It is crucial to say which figure is given.

Example: here is how to calculate the mean, variance and standard deviation of measurements of the widths of the middle fingers of ten people.

Width of middle finger / mm	8, 10, 12, 7, 7, 9, 8, 9, 11, 9

$$\text{Mean} = \frac{8 + 10 + 12 + 7 + 7 + 9 + 8 + 9 + 11 + 9}{10} = \frac{90}{10} = 9$$

Calculating the variance is easier if you put the data into a table:

Width of middle finger / mm	Deviation from mean ($\bar{x} - x$)	Deviation from mean squared ($\bar{x} - x$)^2
8	1	1
10	–1	1
12	–3	9
7	2	4
7	2	4
9	0	0
8	1	1
9	0	0
11	–2	4
9	0	0
$\bar{x} = \frac{90}{10} = 9$	$\Sigma(\bar{x} - x) = 0$	$\Sigma(\bar{x} - x)^2 = 24$

Check your arithmetic.
Make sure this column
adds up to 0.

$$\text{variance} = s^2 = \frac{\Sigma(\bar{x} - x)^2}{10} = \frac{24}{10} = 2.4$$

$$\therefore \quad s = \sqrt{2.4} = 1.55$$

7.6.5 Percentiles and quartiles

Percentiles

A percentile is the value below which a given percentage of observations fall. For example, the 20th percentile is the value below which 20 percent of the observations may be found.

Here are the heart rates, in beats per minute, of 10 people:

46, 54, 55, 62, 66, 66, 69, 71, 74, 95

20% of the heart rates fall at or below 54 so the person with a heart rate of 54 is at the 20th percentile of this group of people.

Quartiles

A quartile represents 25% of a population, so the 1st quartile (Q1) is the 25th percentile. It is the middle number between the smallest number and the median.

Quartile 2 (Q2) is the 50th percentile. It is the median of the data so 50% of the measurements are below Q2.

Quartile 3 (Q3) is the 75th percentile. It is the middle value between the median and the highest value.

Here are 20 heart rates, in beats per minute (bpm):

41 46 53 54 55 55 61 62 65 66 66 66 68 69 71 71 73 74 95 105

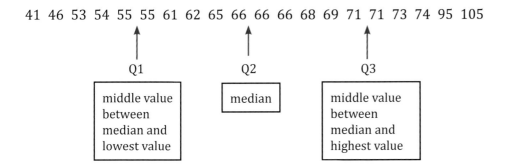

The median value = 66 bpm, with equal numbers of values above and below.

25% of values are at or below 55 bpm so Q1 = 55 and Q3 = 71.

75% of values are at or below 71 bpm so Q3 = 71.

The interquartile range, or 'middle fifty', contains the middle 50% of measurements: IQR = Q3 − Q1 = 71 − 55 = 16.

The diagram shows this data in a box-and-whisker plot. These plots show information about the range of results in a sample, and show how skewed non-normally distributed data are. The median of a population is shown; the 25th and 75th percentiles are the limits of the box, i.e. the interquartile range; the whiskers on the plot show the 10th and 90th percentiles. One of the advantages of this type of plot is that it eliminates the outliers.

With different experiments and data sets, different limits or whiskers may be chosen. With this data set, the 10th percentile is 46 bpm and the 90th percentile is 95 bpm.

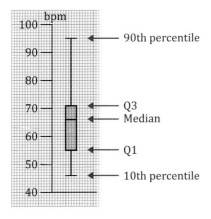

7.7 Making a null hypothesis

There are many types of experiment. Some aim to make a qualitative observation; some aim to make a measurement; some aim to test an idea. The idea is the hypothesis. It is a philosophical notion that it is easier to reject a hypothesis than to accept one. Statistical tests allow the experimenter to reject a hypothesis in order to support the initial idea. The hypothesis to be tested is the null hypothesis. 'Null' means 'nothing' and the null hypothesis states that there is no effect. The 'experimental' or 'alternative hypothesis' states the effect in positive terms so that if the null hypothesis is not supported, the alternative hypothesis is. Bear in mind, statistical tests never prove or disprove; they merely allow you to accept or reject a hypothesis – another philosophical notion.

Here is an example: if an ecologist suspected that increased phosphate ion concentration in river water was correlated with a decrease in the number of stone fly nymphs, the null hypothesis would be: there is no significant correlation between phosphate ion concentration and the number of stone fly nymphs. The alternative hypothesis, however, would state: there is a negative correlation between phosphate ion concentration and the number of stone fly nymphs.

Direction of the hypothesis

In this example, there was no thought that the phosphate ion concentration might increase the number of stone fly nymphs. The alternative hypothesis is therefore described as 'directional', because it suggests the direction of the effect of the phosphate ions. Had the ecologist suspected an increase would happen, then the alternative hypothesis would still have been directional, but in the other direction. The ecologist would have proposed: there is a positive correlation between phosphate ion concentration and the number of stonefly nymphs. A non-directional hypothesis would be phrased as: there is a correlation between phosphate ion concentration and the number of stonefly nymphs.

Genetics and the null hypothesis

In a genetics experiment, the expected ratios are worked out based on Mendelian principles. Certain assumptions are made and the null hypothesis incorporates them:

- There is no difference between the numbers of gametes of each type made by each parent.
- There is no difference in the ability of gametes to fuse with any other gamete.
- There is no difference in the viability of gametes.
- There is no difference in the viability of offspring.
- In the case of dihybrid inheritance, the genes are not linked.

All these assumptions are made when the ratio of offspring is calculated on the basis of Mendelian inheritance. This means that the null hypothesis can be simply stated as 'inheritance is Mendelian'. A consequence of this way of framing the null hypothesis is that, unlike the use of other statistical tests, we seek to accept this null hypothesis, rather than reject it.

>> *Pointer*

Here are the standard symbols for some common terms:

Symbol	For sample	For population
Mean	X	μ
Variance	s^2	σ
Standard deviation	s	σ^2

>> *Pointer*

A statistical test tests a null hypothesis, which states that there is no significant effect.

 7.5

Construct null hypotheses for the following experiments:
(i) Testing the effect of water flow rate on the number of blood worms in a river.
(ii) Testing which of two soft rush populations, one growing at low light intensity and one growing at high light intensity, has longer leaves.
(iii) Testing if the occurrence of free ear lobes and attached ear lobes in a population of people can be explained by the inheritance of two alleles of a single gene, one of which is dominant and the other recessive.

7.8 Level of significance

Imagine your car has broken down and you are forced to spend the night alone in an old dark house that just happens to be nearby. The wind and rain are battering on the roof. Your torch battery is dead and your candles have been blown out by the wind. It is pitch dark, except for glimpses of moon as the clouds scud by. You hear a door slam. As a rational individual, nay a scientist, you think nothing of a chance event and rightly ignore it. But let us say this plays out over 100 nights. How many times would you ignore the door slamming as a chance event, perhaps caused by the wind? On how many nights would it have to slam before you began to consider the prospect of some more sinister explanation?

Statisticians say that if something unexpected happens in 5% of cases or more, then it has a meaning beyond a chance event. So, in the above situation, five nights with a door slamming should make you worry. Four should not. This is an arbitrary figure based on statistical and biological considerations but is well founded and accepted in biology experiments.

Another way of saying this is that if the outcome of an experiment happens in 5% or more of times that the experiment is performed, that outcome has a biological reason and is not due to chance. Statistical tests use the 5% cut-off to allow the experimenter to decide that the null hypothesis has been rejected and that the alternative hypothesis may be accepted. The figure of 5% may be written as a 0.05 probability ($p = 0.05$) that the results are, or are not, due to chance. This is the 5% level of significance that statistical tables show.

In some situations, the 5% level of significance is not considered adequate, and a much smaller value is used, such as 0.001%. This might be the case when evaluating a drug and the side effects with which it is associated.

7.9 Confidence limits

This idea of significance levels is sometimes expressed in terms of confidence limits. If the level of significance is 5% then you are 95% confident that the results are due to a biological phenomenon, and not due to chance. Calculating 95% confidence limits is shown on pages 124–125.

7.10 Degrees of freedom

Degrees of freedom tell us how many factors there are which vary independently within a system. The concept has its origin in n-dimensional geometry but in A Level Biology, the application is statistical tests. Traditionally it has the symbol v, which is the Greek letter 'nu', although it is often abbreviated to the initials df.

There are two situations you are likely to meet in A Level Biology:

a) A sample of independent items where the characteristics of one do not imply anything about the characteristics of any of the others: in a monohybrid cross testing the inheritance of the colour of maize seeds, the seeds could be red or yellow. Let us say that the colour of one is defined, e.g. red. There is only one other possible colour, yellow, so there is one degree of freedom.

» Pointer

If a result is found in at least 5% of trials, a biological mechanism is considered to be the cause.

» Pointer

The biological level of significance is 5%; there is a 0.05 probability that the effect has a biological cause.

quickfire » 7.6

Is there likely to be a biological explanation for these events or not?

(i) 75 seedlings out of 100 offspring of two green tobacco plants being green and 25 being yellow.

(ii) 14 out of 100 students being left handed and 86 out of 100 being right handed.

(iii) 2 dogs out of 50 having three legs and 48 having the normal four.

How many degrees of freedom?
(i) Testing the inheritance of the gene for height in tall and dwarf tomato plants.
(ii) Testing the length of the stipe (stalk) in two populations of the alga *Fucus*, with 45 growing on limestone and 45 on sandstone.
(iii) Testing the inheritance of *Drosophila* genes in a cross between a fly with red eyes and a long body and a fly with white eyes and a short body.

If there are three possible seed colours, red, yellow or white, and the colour of one seed is fixed, e.g. red, there are two other possible choices, yellow and white. There are, therefore, two degrees of freedom.

Consider a dihybrid cross testing the inheritance of genes for guinea pig coat colour and texture. You may expect four phenotypes: black fur + rough coat, black fur + smooth coat, white fur + rough coat and white fur + smooth coat. A black guinea pig may have either a rough or smooth coat. But as the genes segregate independently at meiosis, the four phenotypic classes are still independent of each other. If the phenotype of one is fixed, there are still three other possible classes, giving three degrees of freedom.

The number of degrees of freedom can be described as the number of classes you can choose if one is fixed. So if there are n classes, there are $(n - 1)$ degrees of freedom.

You might use this formulation when testing a hypothesis relating to Mendelian genetics, as on page 143.

b) In testing whether the mean values of two sets of data are different, there may be 30 readings in each set. The number of degrees of freedom for each set is $(30 - 1) = 29$ so the total number of degrees of freedom $= 29 + 29 = 58$. You may meet this if you are comparing the heights of limpet shells on an exposed or sheltered coastline as on page 115.

7.11 Fitting confidence limits to the mean

As shown on page 120, standard deviation is derived from a sample of the population. It is possible to estimate the standard deviation of the whole population, within certain confidence limits. The calculation below will illustrate how to estimate the mean of a whole population, μ, and be 95% certain that the true standard deviation falls within the calculated range. The calculations use Student's t distribution. Student's t test is described later.

Step 1 – Collect the data. Here are the lengths and the mean length of 10 earthworms:

Length / mm										
100	70	50	90	90	60	110	40	90	110	Mean = 81

Step 2 – Square the values to find Σx^2

This step is in the table on the left

Step 3 – Calculate the variance of the sample

Length / mm x	x^2
100	10 000
70	4 900
50	2 500
90	8 100
90	8 100
60	3 600
110	12 100
40	1 600
90	8 100
110	12 100
$\bar{x}$ = 81	Σx^2 = 71 100

Variance of sample, $s^2 = \dfrac{\Sigma x^2}{n} - \bar{x}^2 = \dfrac{71\,100}{10} - 81^2 = 7110 - 6561 = 549$

Step 4 – Estimate the variance of the population

Estimated variance of population, $\sigma^2 = \dfrac{ns^2}{n-1} = \dfrac{10 \times 549}{9} = 610$

Step 5 – Estimate the standard deviation of the population

$\sigma = \sqrt{\text{estimated variance}} = \sqrt{610} = 24.7$

》 Pointer
s = standard deviation of the sample.

σ = standard deviation of the whole population.

Step 6 – Find the level of significance by subtracting confidence level from 100 and converting to decimal

$100 - 95 = 5$

Level of significance, $p = 0.05$

Step 7 – Find df, the number of degrees of freedom

$df = n - 1 = 10 - 1 = 9$.

Step 8 – Use the table of critical values to find the critical value of t_{crit}.

Degrees of freedom	Level of significance			
	0.1	0.05	0.02	0.01
8	1.860	2.306	2.896	3.355
9	1.833	2.262	2.821	3.250
10	1.812	2.228	2.764	3.169
11	1.796	2.201	2.718	3.106

For $n - 1 = 9$ and a significance level of 0.05, $t_{crit} = 2.26$ (3sf).

Step 9 – Calculate the confidence limits:

$$\text{Lower limit} = \bar{x} - \frac{t_{crit}}{\sqrt{n}} \times \sigma = 81 - \frac{2.26 \times 24.7}{\sqrt{10}} = 81 - 17.7 = 63.3 \text{ mm}$$

$$\text{Upper limit} = \bar{x} + \frac{t_{crit}}{\sqrt{n}} \times \sigma = 81 + \frac{2.26 \times 24.7}{\sqrt{10}} = 81 + 17.7 = 98.7 \text{ mm}$$

This means that for the whole population of earthworms, it is 95% likely that the actual mean value will lie between 63.3 mm and 98.7 mm.

Here are the nine steps:

Step	Task
1	Collect the data
2	Square the values to find Σx^2
3	Calculate the variance of the sample
4	Estimate the variance of the population
5	Estimate the standard deviation of the population
6	Find the level of significance
7	Find the number of degrees of freedom
8	Use the table of critical values to find t_{crit}.
9	Calculate the confidence limits

7.12 Ranking

Statistical tests may require you to put the data in order and assign a rank to each value. If each value occurs once, assigning the rank order is straightforward.

Example: the lengths of tunnels made by holly leaf miner larvae

Ten lengths were measured and were 6, 10, 3, 8, 11, 2, 7, 9, 12, 5 mm. They are put in order, as shown in the left-hand column below. Each value is given its number in the order, as shown in the right-hand column.

Lengths in ascending order	Rank
2	1
3	2
5	3
6	4
7	5
8	6
9	7
10	8
11	9
12	10

>> Pointer

Identical values have identical ranks.

If a value appears more than once, then each appearance must have the same rank as it is the same value. The rank order given is the mean of the rank orders the values would have had if they had they been different. For example, if two values were the 1st and 2nd cited, they would have the rank order of $\frac{1+2}{2}$ = 1.5 so they would both have the rank 1.5. Examples of this are shown in the table below.

Example: the body masses of 10 students

Ten students were weighed and were 58, 72, 83, 69, 64, 58, 69, 62, 69, 74 kg. The masses are put in order, as shown in the left-hand column below. Their ranks are calculated and shown in the right-hand column.

Body mass of student / kg	Rank
58	$\frac{1+2}{2} = 1.5$
58	$\frac{1+2}{2} = 1.5$
62	3
64	4
69	$\frac{5+6+7}{3} = 6$
69	$\frac{5+6+7}{3} = 6$
69	$\frac{5+6+7}{3} = 6$
72	8
74	9
83	10

Example: the diameter of ten acorns

The diameters of ten acorns, measured with callipers, were 5, 8, 9, 12, 12, 8, 8, 7, 11, 6 mm.

Diameter of acorn / mm	Rank
5	1
6	2
7	3
8	$\frac{4+5+6}{3} = 5$
8	$\frac{4+5+6}{3} = 5$
8	$\frac{4+5+6}{3} = 5$
9	7
11	8
12	$\frac{9+10}{2} = 9.5$
12	$\frac{9+10}{2} = 9.5$

The rank of the largest value is the same as the number of readings, as long as that value occurs only once. If it occurs more than once, its rank will be an average and therefore lower than the number of readings, as shown in the above example on the right.

7.13 One-tailed and two-tailed tests

When using tables of test statistics, it may be necessary to choose between critical values based on either one- or two-tailed tests. Here is how to decide which to use.

When investigating, it is the null hypothesis that is tested, hypothesising that there will be no significant effect. But if an effect is seen, allowing rejection of the null hypothesis, it could be either positive or negative. An example could be to test if wheat seedlings given extra nitrate ions in their compost produced a different amount of protein from those without. They might produce more but they might produce less and so the test is described as two-tailed, the tails being the two extremes of a normal distribution.

 **7.8**

Rank the following sets of numbers:

(i) 12, 6, 8, 3, 9, 17, 15

(ii) 12, 6, 8, 3, 9, 3, 12

A one-tailed test assumes an effect has a direction; a two-tailed test makes no assumptions about the direction of the effect.

If, however, only one direction of outcome is possible, then the test is one-tailed. An example could be in testing the null hypothesis that alcohol decreases the heart rate of *Daphnia*. Here, no increase is postulated and only the decrease is being tested so the test is one-tailed. A common example is the use of the χ^2 test in genetics. In one direction, if the data are not a good fit for a Mendelian explanation, the null hypothesis is rejected. In the other direction, the fit would be too good, and that is not a situation that has to be considered so a one-tailed test is suitable.

7.14 Correlation

If you measure one factor as a function of another, you may find there is no apparent relationship between them. For example, if you wished to investigate if a correlation exists between the length of a dog's tail and the height of its owner, you may find a short tail may be associated with a tall or a short owner. A scatter graph of the results may look like the diagram below.

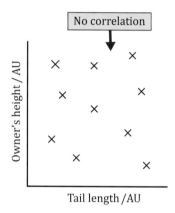

In an experiment testing the relationship between the dog's tail and its body length, you may see that in general, the longer the dog, the longer its tail. The points may lie on a scatter graph as shown in the diagram below.

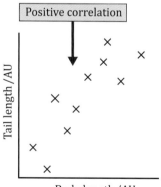

In a positive correlation, as the independent variable increases, the dependent variable increases. In general, lines of best fit are not drawn on correlation graphs. However, if a line of best fit were drawn, depending on the nature of the experiment, the line describing a positive correlation may go through the origin.

The graph on page 130 shows an experiment testing the effect of light intensity on the internode length of nettles. As the light intensity increases, the internode length decreases. This is a negative correlation.

In a negative correlation, as the independent variable increases, the dependent variable decreases.

The degree of correlation is described a 'correlation coefficient', which quantifies how good the correlation is. It can range between −1, for a perfect negative correlation, and +1, for a perfect positive correlation.

Each point on the scatter graph is a single piece of information. There are no replicate readings so no range bars can be drawn. However, the reliability of the sample can be assessed by judging how close points lie to the imagined line of best fit. The closer to the line they are, the more reliable the data set. It is important to recognise that just because two factors are correlated, it does not mean that one necessarily causes the other. In the example above, the long body of a dog does not cause it to have a long tail. It may be that the same thing causes both, or it may be that they are entirely unrelated and the correlation is a matter of chance.

Negative correlation

Internode length /AU

Light intensity /AU

7.15 Choosing a statistical test

Statistical tests are designed for specific uses and so it is important to choose the test that suits the data, the sample size, the distribution and the direction of the alternative hypothesis. Statistics text books may show you flow charts allowing you to identify which of the many available tests is appropriate. At A Level Biology, however, only a small number of tests are commonly used. You would never be expected to remember the formulae involved and if you were performing a test in an examination, all the tables and formulae would be provided.

In the context of statistical testing, data can be considered in three different ways:

a) Categorical data occur in categories such as seeds being smooth or wrinkled or people having their left or right side dominant. It is sometimes called nominal data.

b) Ordinal data describes variables that cannot be measured accurately, but can be ranked in order, such as giving a point score to describe how dark an indicator is.

c) Interval data are accurate measurements, such as plant height.

The types of samples may also be described in different ways:

a) Matched samples are in pairs, such as the area of lichen cover on the north and south sides of grave stones. The area in the north side of a single stone is matched with the area on the south side.

b) The term 'paired' is used if the value of the dependent variable is being tested against the independent variable to see if they are correlated, such as the number of water shrimp in a given depth of river. They are not 'matched' as they are different factors but they are 'paired' as each number belongs with a particular depth.

c) Independent samples are selected at random from the entire population and so do not belong in pairs, such as the area of bramble leaves growing at low and high light intensities.

The table below shows how you may choose between the tests, and shows an example of an investigation in which you might use each of them.

>> *Pointer*

Can you predict an outcome? Use the χ^2 test.

>> *Pointer*

Are you testing a non-linear correlation? Use the Spearman rank correlation test.

>> *Pointer*

Are you testing a linear correlation? Use the Pearson linear correlation test.

>> *Pointer*

Do you have data with a normal distribution? Use the *t* test.

>> *Pointer*

Do you have data with a non-normal distribution? Use the Mann-Whitney U test.

Test	Data type	Sample type	Distribution	Sample size	Examples of use
Spearman rank correlation	Ordinal	Paired	Non-normal distribution; non-linear	≥10 pairs. 5–9 pairs can be used, but test is less reliable	Testing if eye colour and hair colour are correlated
	Interval				Testing for a correlation between concentration of lead ions in the growth medium and the length of roots of newly germinated seeds
Pearson linear correlation (PMCC)	Interval	Paired	Normal distribution; linear	≥10 pairs. 5–9 pairs can be used, but test is less reliable	Correlating haemoglobin level with packed cell volume
Mann-Whitney U test	Ordinal	Independent		< 30	Resazurin test – judging if the colour of the dye is white, pink, purple or blue as an estimate of bacterial growth
	Interval	Independent	Non-normal	< 30	Comparing the width of ground ivy leaves under a tree and in the middle of a field
t test	Interval	Independent	Normal	< 30	Comparing the heights of limpets on exposed and sheltered shores
	Interval	Matched	Normal	< 30	Comparing stomatal counts on the upper and lower surfaces of leaves
χ^2	Categorical			expected values ≥5 in > 80% cells of table	Experiments where an outcome can be predicted, e.g. testing for Mendelian inheritance

7.16 The Spearman rank correlation test

When people say, for short, that they are using the 'Spearman rank' test, it suggests they do not understand that it is a rank correlation test. The ranks are being correlated and Spearman is not being ranked. It is best to give the test its full name.

Here is a worked example: this experiment is testing to see if there is a correlation between the length of the fourth internode of bramble and the light intensity where it is growing.

Step 1 – Construct a null hypothesis: there is no significant correlation between light intensity and the length of the fourth internode of bramble.

Step 2 – Collect the data

Bramble

Sample number	Light intensity / lux	Internode length / mm
1	1200	18
2	800	23
3	6100	13
4	2100	17
5	300	25
6	4800	13
7	6900	12
8	9300	5
9	3600	15
10	8100	8

Step 3 – Rank the light intensities. The column is headed R_1. When you re-write them in order, make sure that each light intensity reading is still paired in the table with its related internode length.

Step 4 – Rank the internode lengths. The column is headed R_2 Remember that if two values share a rank, their rank number is the mean of the two ranks that would have been occupied had they been different.

Light intensity / lux	Rank R_1	Internode length / mm	Rank R_2
300	1	25	10
800	2	23	9
1200	3	18	8
2100	4	17	7
3600	5	15	6
4800	6	13	4.5
6100	7	13	4.5
6900	8	12	3
8100	9	8	2
9300	10	5	1

» Pointer

$\Sigma d = 0$.

Step 5 – Calculate the difference between the ranks, *d*. It does not matter which you take from which, as long as you do it the same way for each pair. In this example, R_2 is taken away from R_1. Some values of *d* will be positive and some will be negative. As a check, if you add the differences, they come to zero, i.e. $\Sigma d = 0$. If yours do not, it is likely that they have been ranked incorrectly.

Light intensity / lux	Rank R_1	Internode length / mm	Rank R_2	Difference between ranks $R_1 - R_2 = d$
300	1	25	10	−9
800	2	23	9	−7
1200	3	18	8	−5
2100	4	17	7	−3
3600	5	15	6	−1
4800	6	13	4.5	1.5
6100	7	13	4.5	2.5
6900	8	12	3	5
8100	9	8	2	7
9300	10	5	1	9

The sum of this column is zero. $\Sigma d = 0$

Step 6 – Square the differences, and add them to get Σd^2

Light intensity / lux	Rank R_1	Internode length / mm	Rank R_2	Difference between ranks $R_1 - R_2 = d$	d^2
300	1	25	10	−9	81
800	2	23	9	−7	49
1200	3	18	8	−5	25
2100	4	17	7	−3	9
3600	5	15	6	−1	1
4800	6	13	4.5	1.5	2.25
6100	7	13	4.5	2.5	6.25
6900	8	12	3	5	25
8100	9	8	2	7	49
9300	10	5	1	9	81
Σd^2					328.5

Step 7 – Calculate the test statistic r_s using the equation:

$$r_s = 1 - \frac{6\Sigma d^2}{n(n^2 - 1)} \text{ where } n = \text{number of pairs} = 10.$$

$$r_s = 1 - \frac{6 \times 328.5}{10(100 - 1)} = 1 - \frac{1971}{990} = 1 - 1.99 = -0.99$$

Step 8 – Interpret the value of r_s

8a) Look at the sign. If r_s is positive, the correlation is positive, and the dependent variable increases as the independent variable increases. If r_s is negative, the correlation is negative and the dependent variable decreases as the independent variable increases.

The sign only signals the direction of the correlation. It does not say if the correlation is linear or not, and this test does not test for linearity. It can therefore be used whether or not the line of best fit would be a straight line.

In this example, $r_s = -0.99$. It is negative and so the correlation is negative, in other words, as the light intensity increases, the length of the fourth internode of bramble decreases.

8b) Test for significance at the 0.05 level of significance. A scatter graph of these readings would show a steep gradient, suggesting a strong correlation. As explained on pages 128–129, the correlation coefficient lies between +1 and −1. So −0.99 sounds like a very strong correlation but without comparing the calculated value of r_s with a critical value derived from tables, the strength of the correlation cannot be ascertained.

This is a two-tailed test because the direction of the correlation was not predicted. The table below is an extract from a table of critical values for this test. It shows the critical values for r_s for different numbers of pairs of results, for different levels of significance.

Number of pairs	Level of significance				
	0.20	0.10	0.05	0.02	0.01
8	0.5476	0.6429	0.7381	0.8095	0.8571
9	0.4833	0.6000	0.6833	0.7667	0.8167
10	0.4424	0.5636	0.6485	0.7333	0.7818
11	0.4182	0.5273	0.6091	0.7000	0.7545

For this example, for 10 pairs of readings, $n = 10$ and the level of significance is 0.05.

From the table, the critical value is 0.6485. This can be written $r_{s\,crit} = 0.6485$. It describes how closely the data must approximate to a line that would indicate correlation. What follows ignores the sign of the calculated value of r_s, $r_{s\,calc}$. If it is equal to or higher than this critical value, then the experimental data are closer than the putative test data and the correlation is significant. If the calculated value is smaller, it is not close enough to the theoretical data and so there is no correlation. In this case the calculated value is numerically higher (0.99) than the critical value (0.6485). This can be written as $r_{s\,calc} > r_{s\,crit}$, where '>' means 'is greater than'.

Step 9 – Consider the null hypothesis. Since $r_{s\,calc} > r_{s\,crit}$ there is a significant correlation so the null hypothesis is rejected.

Step 10 – Formulate the conclusion. The final sentence of any statistical test must contain four elements:

a) Compare the calculated and critical values for the test statistic.

b) Accept or reject the null hypothesis.

c) State level of significance.

7.9

Reformulate the conclusion if the calculated value $r_s = 0.4829$.

d) Say what all that means.

So in this example, the conclusion would read:

> $r_{s\,calc} = -0.99$ and $r_{s\,crit} = 0.6485$.
>
> Ignoring the sign, $r_{s\,calc} > r_{s\,crit}$ so the null hypothesis is rejected at the 0.05 level of significance. There is a negative correlation between light intensity and the length of the fourth internode of bramble.

It is important to remember that just because two factors are correlated with each other, it does not mean that one causes the other. They may both be caused by other factors, either the same or different. Correlation implies nothing about the origin of or reason for the observation.

Here are the ten steps:

Step	Task
1	Construct a null hypothesis
2	Collect the data
3	Rank the values of independent variable
4	Rank the values of the dependent variable
5	Calculate the difference between the ranks, d
6	Square the differences, and add them to get Σd^2
7	Calculate the test statistic r_s
8a	Interpret the value of r_s: look at the sign
8b	Interpret the value of r_s: test for significance at the 0.05 level of significance
9	Consider the null hypothesis
10	Formulate the conclusion

7.17 The Pearson linear coefficient test

This is also called the product-moment correlation coefficient (PMCC) or the Pearson product-moment correlation coefficient test (PPMCC). It is used when you wish to test the strength of a correlation that is linear. The value of r lies between -1 and $+1$, where $r = -1$ shows a perfect negative correlation, $r = 1$ shows a perfect positive correlation and $r = 0$ shows no correlation. This does not mean that there is no relationship, only that if there is a relationship, it is not linear.

The correlation coefficient, r, gives information about how close all the points are to a presumed straight line of best fit but, unlike the Spearman rank correlation coefficient, does not give information about the gradient of the curve. To ascertain how strong the correlation is, the coefficient of determination, r^2 is used

Step 1 – Construct a null hypothesis: there is no significant correlation between the time since germination of a pollen grain and the length of pollen tube.

Step 2 – Collect the data.

This is shown in the table on the right.

Time / minutes x	Pollen tube length / µm y
0	0
30	0
60	5
90	10
120	18
150	22
180	25
210	29
240	34
270	40

Step 3 – Find xy and square each value of x and y

x	x^2	y	y^2	xy
0	0	0	0	0
30	900	0	0	0
60	3600	5	25	300
90	8100	10	100	900
120	14400	18	324	2160
150	22500	22	484	3300
180	32400	25	625	4500
210	44100	29	841	6090
240	57600	34	1156	8160
270	72900	40	1600	10800

Step 4 – Sum columns

x	x^2	y	y^2	xy
0	0	0	0	0
30	900	0	0	0
60	3600	5	25	300
90	8100	10	100	900
120	14400	18	324	2160
150	22500	22	484	3300
180	32400	25	625	4500
210	44100	29	841	6090
240	57600	34	1156	8160
270	72900	40	1600	10800
$\Sigma x = 1350$	$\Sigma x^2 = 256500$	$\Sigma y = 183$	$\Sigma y^2 = 5155$	$\Sigma xy = 36210$

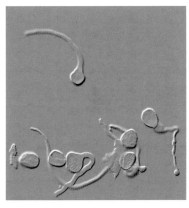

Germinating pollen grains

Step 5 – Substitute into the equation: $r = \dfrac{n(\Sigma xy) - (\Sigma x)(\Sigma y)}{\sqrt{[n\Sigma x^2 - (\Sigma x)^2][n\Sigma y^2 - (\Sigma y)^2]}}$

where r = product-moment correlation coefficient and n = number of samples.

$$r = \frac{10(36\,210) - (1350)(183)}{\sqrt{[10 \times 256\,500 - (1350)^2][10 \times 5155 - (183)^2]}}$$

$$= \frac{362\,100 - 247\,050}{\sqrt{[2\,565\,000 - 1\,822\,500][51\,550 - 33\,489]}}$$

$$= \frac{115\,050}{\sqrt{[742\,500][18\,061]}}$$

$$= \frac{115\,050}{115\,803}$$

$$= 0.99$$

Step 6 – Interpret r:

6a – the sign. If r is positive, the correlation is positive and if r is negative, the correlation is negative. In this case, it is positive.

6b – the significance. In the example given, a one-tailed test is used, as the direction of the correlation is assumed, i.e. a pollen tube gets longer. Using the extract of the table of critical values given below, for 10 pairs of readings and a level of significance of 0.05, $r_{crit} = 0.5494$.

Number of pairs	Level of significance				
	0.10	0.05	0.025	0.01	0.005
8	0.5067	0.6215	0.7067	0.7887	0.8343
9	0.4716	0.5822	0.6664	0.7498	0.7977
10	0.4428	0.5494	0.6319	0.7155	0.7646
11	0.4187	0.5214	0.6021	0.6851	0.7348

Step 7 – Consider the null hypothesis: Since $r_{s\,calc} > r_{s\,crit}$ there is a significant correlation so the null hypothesis is rejected.

Step 8 – Calculate the coefficient of determination. To determine the strength of the correlation, the value of r must be squared to give the coefficient of correlation. This shows how strong the correlation is. The table below shows how this value is interpreted:

Coefficient of determination, r^2	Strength of correlation between variables
0	None
0.2	Low
0.3	Weak
0.3 – 0.7	Moderate
0.7 – 1	Strong

In this case, $r_{calc} = 0.99$ so the correlation is strong.

Step 9 – Formulate the conclusion. The final sentence of any statistical test must contain four elements:

a) Compare the calculated and critical values for the test statistic.

b) Accept or reject the null hypothesis.

c) State level of significance.

d) Say what all that means.

So in this example, the conclusion would read:

$r_{s\,calc} = 0.99$ and $r_{s\,crit} = 0.5494$

$r_{s\,calc} > r_{s\,crit}$ so the null hypothesis is rejected at the 0.05 level of significance. There is a correlation between the time since germination and the length of the pollen tube.

$r_{s\,calc}$ is positive so the correlation is positive.

$r_{s\,calc} = 0.99$ $r_{s\,calc} = 0.99^2 = 0.98$ the correlation is strong.

Here are the nine steps:

Step	Task
1	Construct a null hypothesis
2	Collect the data
3	Find xy and square each value of x and y
4	Sum columns
5	Substitute into the equation to find r
6a	Interpret the sign
6b	Assess for significance at the 0.05 level of significance
7	Consider the null hypothesis
8	Calculate the coefficient of determination
9	Formulate the conclusion

7.18 The Mann-Whitney U test

This test is used when you have categorical data or a sample of interval data that is not normally distributed. It compares median values. Here is a worked example: this experiment is to test if there is a difference in numbers of water shrimp in rivers with sandy or stony substrate.

Step 1 – Construct a null hypothesis: there is no significant difference between the median number of water shrimp in rivers with sandy or stony substrate.

Step 2 – Collect the data

Substrate	Number of water shrimp in kick sample							
Sandy	3	4	2	6	0	0	1	2
Stony	8	5	6	9	10	9	12	13

Step 3 – Rank the data. For this test, the data are ranked as if they are one sample, so with 8 readings in each site, there are 8 × 2 = 16 ranks altogether. It is easiest if the results table is re-written so that ascending counts and overlap of readings in the two samples can be recognised:

Substrate	Number of water shrimp in kick sample													
Sandy	0	0	1	2	2	3	4		6					
Stony								5		6	8	9	9	10	12	13

The entire data are ranked as one set of figures. Remember that when more than one value appears, they all have the same rank, which is the mean of the ranks they would occupy if they were not the same.

Substrate	Water shrimp in kick sample														
Sandy	0	0	1	2	2	3	4		6							
Rank $_{sandy}$	1.5	1.5	3	4.5	4.5	6	7		9.5							
Stony								5		6	8	9	9	10	12	13
Rank $_{stony}$								8		9.5	11	12.5	12.5	14	15	16

Step 4 – Sum the ranks

For the sandy substrate, the sum of ranks,

$$\Sigma R_1 = 1.5 + 1.5 + 3 + 4.5 + 4.5 + 6 + 7 + 9.5 = 37.5$$

For the stony substrate, the sum of ranks,

$$\Sigma R_2 = 8 + 9.5 + 11 + 12.5 + 12.5 + 14 + 15 + 16 = 98.5$$

Step 5 – Calculate the test statistics, U_1 for the sandy site and U_2 for the stony site using the equation shown, where n_1 = sample size in the sandy site and n_2 = sample size in the stony site.

$U_1 = n_1 \times n_2 + \frac{1}{2}n_2(n_2 + 1) - R_2$	$U_2 = n_1 \times n_2 + \frac{1}{2}n_1(n_1 + 1) - R_1$
$= 8 \times 8 \quad + \frac{1}{2} \times 8(8 + 1) - 98.5$	$= 8 \times 8 \quad + \frac{1}{2} \times 8(8 + 1) - 37.5$
$= 64 \quad + \quad 36 \quad - 98.5$	$= 64 \quad + \quad 36 \quad - 37.5$
$= 1.5$	$= 62.5$

> **Pointer**
>
> $U_1 + U_2 = n_1 \times n_2$.

Step 6 – Check your arithmetic. You can check your arithmetic at this stage, because if you are correct, $U_1 + U_2 = n_1 \times n_2$. If they are not equal, there is probably an error in the ranking.

In this example, $1.5 + 62.5 = 8 \times 8 = 64$.

Step 7 – Compare calculated test statistic with critical value

U can be thought of as a measure of similarity between the median values for the two samples. As we wish to reject the hypothesis that they are similar, we would like our calculated value to be lower than the critical value, i.e. less similar, which is the same as more different. So if the calculated value is the same or lower than the critical value, the null hypothesis can be rejected.

On the left is an extract of the table for critical values of U for this test for a significance level of 0.05.

		n_2			
		7	**8**	**9**	**10**
	7	8	10	12	14
n_1	**8**		13	15	17
	9			17	20
	10				23

The lower value of U is used to compare with the critical value. In this example, the lower value = 1.5.

Step 8 – Consider the null hypothesis

For $n_1 = n_2 = 8$, $U_{crit} = 13$. $U_{calc} < U_{crit}$, where '<' means 'is lower than', so the null hypothesis can be rejected.

Step 9 – Formulate the conclusion. As stated above, the final sentence of any statistical test must contain four elements:

a) Compare the calculated and critical values for the test statistic.

b) Accept or reject the null hypothesis.

c) State level of significance.

d) Say what all that means.

So in this example, the conclusion would read:

> The lower $U_{calc} = 1.5$ and $U_{crit} = 13$.
>
> $U_{calc} < U_{crit}$ so the null hypothesis is rejected at the 0.05 level of significance. The median number of water shrimp on a sandy substrate is lower than that on a stony substrate.

> **quickfire** >> 7.10
>
> Reformulate the conclusion if $U_1 = 20$ and $U_2 = 44$.

Here are the nine steps:

Step	Task
1	Construct a null hypothesis
2	Collect the data
3	Rank the data as one set
4	Sum the ranks to find ΣR_1 and ΣR_2
5	Calculate the test statistics, U_1 and U_2
6	Check your arithmetic
7	Compare lower calculated U with critical value
8	Consider the null hypothesis
9	Formulate the conclusion

7.19 The *t* test

You would use the *t* test to compare the means of two samples of interval data with up to 30 values normally distributed in each set. There are different formulations for calculating the test statistic, *t*, but in an examination, you would be given the appropriate equation. The formula used depends on whether the sets of data are matched or unmatched, as explained on page 129, and whether the sample sizes and variance for the two sets are the same. In A Level Biology, a common use of this test involves comparing populations, such as of blood cells or of animal or plant populations in a fieldwork experiment. In such cases, the samples may or may not be equal and as the variance is unknown, it cannot be assumed to be the same for the two sets of data. The worked example below uses the equation appropriate for these conditions.

Worked example: the experiment is to test if there is a difference between the numbers of mayfly nymphs in streams with a high or low concentration of dissolved oxygen.

Step 1 – Construct the null hypothesis: there is no significant difference between the mean number of mayfly nymphs in streams with high or low concentration of dissolved oxygen.

Step 2 – Collect the data

This is shown in the upper table on the right.

Step 3 – Calculate the mean for each set: mean = $\bar{x} = \dfrac{\text{sum of values}}{\text{number of values}} = \dfrac{\Sigma x}{10}$

This is shown in the lower table on the right.

Step 4 – Calculate the variance for each set: variance = $s^2 = \dfrac{\Sigma(\bar{x} - x)^2}{10}$

This was described on page 120 and is shown again on page 140. Here is a summary:

a) Calculate the deviation of each value from the mean.

b) Check your arithmetic and make sure that $\Sigma d_1 = 0$ and $\Sigma d_2 = 0$.

c) Calculate the square of the deviation from the mean.

d) Add the square of the deviations.

e) Divide by the number of values.

Number of mayfly nymphs in kick sample at dissolved oxygen concentration	
High	**Low**
15	9
10	5
8	7
23	2
17	7
17	8
20	0
12	10
9	6
13	5

Number of mayfly nymphs in kick sample at dissolved oxygen concentration		
	High	**Low**
	15	9
	10	5
	8	7
	23	2
	17	7
	17	8
	20	0
	12	10
	9	6
	13	5
Mean	$\dfrac{144}{10} = 14.4$	$\dfrac{59}{10} = 5.9$

Mayfly nymphs in kick sample at high dissolved oxygen concentration			Mayfly nymphs in kick sample at low dissolved oxygen concentration		
Number	Deviation from mean $(\bar{x} - x)$	Deviation from mean² $(\bar{x} - x)^2$	Number	Deviation from mean $(\bar{x} - x)$	Deviation from mean² $(\bar{x} - x)^2$
15	−0.6	0.36	9	−3.1	9.61
10	4.4	19.36	5	0.9	0.81
8	6.4	40.96	7	−1.1	1.21
23	−8.6	73.96	2	3.9	15.21
17	−2.6	6.76	7	−1.1	1.21
17	−2.6	6.76	8	−2.1	4.41
20	−5.6	31.36	0	5.9	34.81
12	2.4	5.76	10	−4.1	16.81
9	5.4	29.16	6	−0.1	0.01
13	1.4	1.96	5	0.9	0.81
$\bar{x} = 14.4$	$\Sigma d_1 = 0$	$\Sigma(\bar{x} - x)^2 = 216.40$	$x = 5.9$	$\Sigma d_2 = 0$	$\Sigma(\bar{x} - x)^2 = 84.90$

$$s^2 = \frac{\Sigma(\bar{x} - x)^2}{10} = \frac{216.40}{10} = 21.640 \qquad s^2 = \frac{\Sigma(\bar{x} - x)^2}{10} = \frac{84.90}{10} = 8.490$$

Step 5 – Calculate the test statistic t:

Here the equation for t is $t = \dfrac{(\bar{x}_{high} - \bar{x}_{low})}{\sqrt{\dfrac{s^2_{high}}{n} + \dfrac{s^2_{low}}{n}}}$

$$t = \frac{(14.4 - 5.9)}{\sqrt{\dfrac{21.640}{10} + \dfrac{8.490}{10}}} = \frac{8.5}{\sqrt{2.164 + 0.849}} = \frac{8.5}{\sqrt{3.013}} = \frac{8.5}{1.736} = 4.896$$

Step 6 – Calculate the number of degrees of freedom: this test uses both sets of data and so the total number of degrees of freedom is the sum of the number of degrees of freedom for each set.

df = (10 − 1) + (10 − 1) = 18

Step 7 – Compare calculated value for t with critical value: the calculated value of t must be compared with a critical value of t to test if the means of the two sets of data are different at the 0.05 level of significance. It may help to think of t as a measure of difference between the means and so we want to exceed a certain value of t, which would say that our results are even more different. That way we may reject the null hypothesis, which says there is no difference.

Here is an extract from the table of critical values for t:

Degrees of freedom	Level of significance		
	0.1	0.05	0.02
17	1.740	2.110	2.567
18	1.734	2.101	2.552
19	1.729	2.093	2.539

The required critical value is for 18 degrees of freedom and 0.05 level of significance.

Thus t_{crit} = 2.101.

From the calculations in Step 5, t_{calc} = 4.891.

$$t_{calc} > t_{crit}$$

Step 8 – Consider the null hypothesis: If t_{calc} is equal to or greater than t_{crit} the null hypothesis can be rejected.

Step 9 – Formulate the conclusion. As stated above, the final sentence of any statistical test must contain four elements:

a) Compare the calculated and critical values for the test statistic.

b) Accept or reject the null hypothesis.

c) State level of significance.

d) Say what all that means.

So in this example, the conclusion would read:

> t_{calc} = 4.891 and t_{crit} = 2.101.
>
> $t_{calc} > t_{crit}$ so the null hypothesis is rejected at the 0.05 level of significance. The mean number of mayfly nymphs in water with a high concentration of dissolved oxygen is greater than in a low concentration.

 7.11

Reformulate the conclusion if the calculated value of t, t_{calc} = 1.375.

Here are the nine steps:

Step	Task
1	Construct the null hypothesis
2	Collect the data
3	Calculate the mean for each set of data
4	Calculate the variance for each set
5	Calculate the test statistic t
6	Calculate the number of degrees of freedom
7	Compare calculated value for t with critical value
8	Consider the null hypothesis
9	Formulate the conclusion

7.20 The χ^2 test

This test is used when you can put your data into categories, so it is used with categorical data. It requires you to make a prediction about what data you expect. That expectation is compared with the observed data to confirm whether the observations are likely to have arisen by chance or whether a biological mechanism might explain them.

A common use in A Level Biology is in analysing data derived from genetic crosses, and a worked example of that is given here. This example will test the outcome of a cross between two maize plants which produce yellow seeds and have the genotype Yy, where Y is the allele producing yellow seeds and y is the allele producing white seeds.

Yellow maize seeds

White maize seeds

Step 1 – Construct the null hypothesis: the null hypothesis for genetics crosses was explained on page 122. The null hypothesis is that inheritance is Mendelian, producing offspring with the ratio of 3 producing yellow seeds : 1 producing white seeds.

Step 2 – Collect the data:

	Yellow seeds	White seeds
Number of offspring	80	20

Step 3 – Calculate expected values based on the ratio 3 : 1 and tabulate them:

Yellow seeds are expected to be three quarters of the total.

Total = 80 + 20 = 100

number of yellow seeds = $\frac{3}{4} \times 100 = 75$

White seeds are expected to be one quarter of the total.

Total = 80 + 20 = 100

number of white seeds = $\frac{1}{4} \times 100 = 25$

		Number of seeds	
		Observed	Expected
Number of offspring	**Yellow seeds**	80	75
	White seeds	20	25

Step 4 – Calculate the difference between the observed (O) and expected (E) values:

		Number of seeds		
		Observed	Expected	$O - E$
Number of offspring	**Yellow seeds**	80	75	5
	White seeds	20	25	−5

Step 5 – Calculate $(O - E)^2$:

		Number of seeds		$O - E$	$(O - E)^2$
		Observed	Expected		
Number of offspring	**Yellow seeds**	80	75	5	25
	White seeds	20	25	−5	25

Step 6 – Divide each value of $(O - E)^2$ by E:

		Number of seeds		$O - E$	$(O - E)^2$	$\dfrac{(O - E)^2}{E}$
		Observed	Expected			
Number of offspring	Yellow seeds	80	75	5	25	0.33
	White seeds	20	25	−5	25	1.00

Step 7 – Add up all values for $\dfrac{(O - E)^2}{E}$ to calculate $\sum \dfrac{(O - E)^2}{E} = \chi^2$:

		Number of seeds		$O - E$	$(O - E)^2$	$\dfrac{(O - E)^2}{E}$
		Observed	Expected			
Number of offspring	Yellow seeds	80	75	5	25	0.33
	White seeds	20	25	−5	25	1.00
					$\chi^2 = \sum \dfrac{(O - E)^2}{E}$	1.33

Step 8 – Find the number of degrees of freedom: as there are two independent classes, yellow and white, df = (number of classes − 1) = (2 − 1) = 1

Step 9 – Find the critical value for χ^2: the calculated value of χ^2 is compared with the critical value, derived from tables. An extract of the relevant table is shown here and the critical value is where df = 1 and the level of significance is 0.05.

Degrees of freedom	Level of significance				
	0.1	0.05	0.01	0.005	0.001
1	2.71	3.84	6.63	7.88	10.82
2	4.61	5.99	9.21	10.60	13.81
3	6.25	7.81	11.34	12.84	16.27

From this table, $\chi^2_{crit} = 3.84$.

Step 10 – Compare calculated value for χ^2 with critical value:

$\chi^2_{crit} = 3.84$ and $\chi^2_{calc} = 1.33$ so $\chi^2_{crit} > \chi^2_{calc}$

Step 11 – Consider the null hypothesis:

If the calculated value of χ^2 is equal to or greater than the value for 0.05 level of significance, the inheritance of this gene is less than 0.05 (=5%) likely to be due to chance. It is at least 95% likely to have another explanation so the null hypothesis can be accepted at the 0.05 level of significance.

Step 12 – Formulate the conclusion: as mentioned above, the final sentence of any statistical test must contain four elements:

a) Compare the calculated and critical values for the test statistic.

b) Accept or reject the null hypothesis.

c) State level of significance.

d) Say what all that means.

>> *Pointer*

Don't forget to add 'any deviation from the ideal ratio is due to chance'.

quickfire >> 7.12

Reformulate the conclusion if the calculated value of χ^2 were 4.68.

So in this example, the conclusion would read:

> χ^2_{crit} = 3.84 and χ^2_{calc} = 1.00.
>
> $\chi^2_{crit} > \chi^2_{calc}$ so the null hypothesis is accepted at the 0.05 level. Inheritance is Mendelian and any deviation from the ideal ratio is due to chance.

Here are the 12 steps:

Step	Task
1	Construct the null hypothesis
2	Collect the data
3	Calculate expected values
4	Calculate the difference between the observed (O) and expected (E) values
5	Calculate $(O - E)^2$
6	Divide each value of $(O - E)^2$ by E
7	Add up all values for $\dfrac{(O - E)^2}{E}$ to calculate χ^2
8	Find degrees of freedom
9	Find the critical value for χ^2
10	Compare calculated value for χ^2 with critical value
11	Consider the null hypothesis
12	Formulate a conclusion

A Fuji apple

Degrees of freedom and the χ^2 test

When the categories of data are in independent classes, as in the above example, df = (number of classes −1). However, if the classes are not independent, the number of degrees of freedom must take this into account. Consider an experiment that tested if Braeburn apple trees and Fuji apple trees produced equal numbers of predominately red or green fruit. The grid for recording the counts would look like this:

Apple variety	Apple colour	
	Red	**Green**
Braeburn		
Fuji		

Braeburns can be red or green. Fujis can be red or green. So the four classes are not independent of each other. In this case,

df = (rows − 1) × (columns − 1) = 1 × 1 = 1.

A Braeburn apple

Test yourself 7

① It has been suggested that the percentage area cover of heather (*Erica tetralix*) is related to the pH of the soil in which it grows and that an increase in pH above 4.5 is detrimental to its growth. The following estimates were collected using a 10 × 10 gridded square quadrat with sides of 0.5m:

Soil pH	% area cover of heather
4.5	100
5.0	97
5.5	92
6.0	92
6.5	76
7.0	83
7.5	70
8.0	61
8.5	66
9.0	64

a) Explain why the Spearman rank correlation test is appropriate for analysing these data.

b) What null hypothesis would be suitable for this test?

c) Test the null hypothesis using the equation $r_s = 1 - \dfrac{6\Sigma d^2}{n(n^2 - 1)}$ and the following extract from the table of critical values:

Number of pairs	Level of significance				
	0.20	0.10	0.05	0.02	0.01
8	0.5476	0.6429	0.7381	0.8095	0.8571
9	0.4833	0.6000	0.6833	0.7667	0.8167
10	0.4424	0.5636	0.6485	0.7333	0.7818
11	0.4182	0.5273	0.6091	0.7000	0.7545

② The following measurements were collected for the growth of a callus of tobacco cells in tissue culture. Devise a null hypothesis and use the Pearson product-moment correlation coefficient to determine whether the growth is correlated with its time in culture.

Time / days x	Callus diameter / mm y
0	2
2	2
4	3
6	14
8	6
10	8
12	12
14	13
16	15
18	16

The correlation may be tested using the table of critical values below:

Number of pairs	Level of significance				
	0.10	0.05	0.025	0.01	0.005
8	0.5067	0.6215	0.7067	0.7887	0.8343
9	0.4716	0.5822	0.6664	0.7498	0.7977
10	0.4428	0.5494	0.6319	0.7155	0.7646
11	0.4187	0.5214	0.6021	0.6851	0.7348

❸ A student collected damselfly larvae at two sites on a river. The table shows the counts:

a) Explain why the Mann-Whitney U test is a suitable statistical test to compare the numbers at the two sites.

Site number	Number of damselfly larvae					
1	10	9	15	16	21	9
2	2	3	5	12	6	8

b) Give a null hypothesis to compare the data.

c) Apply the Mann-Whitney U test to test the null hypothesis, using the equations

$$U_1 = n_1 \times n_2 + \tfrac{1}{2}n_2(n_2 + 1) - \Sigma R_2$$

$$U_2 = n_1 \times n_2 + \tfrac{1}{2}n_1(n_1 + 1) - \Sigma R_1$$

and this extract from the table of critical values:

❹ The number of lymphocytes was counted in samples of blood of the same volume taken from a person who had a viral infection (A) and a healthy person (B). A null hypothesis that there was no significant difference between the numbers of lymphocytes in the two subjects was proposed.

		n_2				
		4	5	6	7	8
n_1	4	0	1	2	3	4
	5	1	2	3	5	6
	6	2	3	5	6	8
	7	3	5	6	8	10
	8	4	6	8	10	13

Use the t test to test this hypothesis, using the equation $t = \dfrac{(\bar{x}_A - \bar{x}_B)}{\sqrt{\dfrac{s^2_A}{n_A} + \dfrac{s^2_B}{n_B}}}$

where the variance for sample A, $s^2_A = 227.36$ and the variance for sample B, $s^2_B = 173.45$. n_A is the number in sample A and n_B is the number in sample B.

Number of lymphocytes	
Sample A	Sample B
176	150
189	167
165	143
157	189
187	158
196	154
168	147
176	168
165	155
196	187

An extract from the table of critical values for t is given below:

Degrees of freedom	Level of significance		
	0.1	0.05	0.01
16	1.746	2.120	2.921
17	1.740	2.110	2.898
18	1.734	2.101	2.878
19	1.729	2.093	2.861
20	1.725	2.086	2.845

5. An experiment was carried out with tomato plants in which heterozygous plants with purple stems were allowed to interbreed. Of the 480 seedlings produced, 346 had purple stems and 134 had green stems. Use the χ^2 test to test if this demonstrates Mendelian inheritance.

The equation to apply is $\chi^2 = \sum \dfrac{(O - E)^2}{E}$.

The table shows critical values for χ^2.

Degrees of freedom	Level of significance			
	0.5	0.1	0.05	0.01
1	0.455	2.706	3.841	6.635
2	1.386	4.605	5.991	9.210
3	2.366	6.251	7.815	11.341

Quickfire answers

1 Numbers

1.1 $17 + 17 = 34$

1.2 $14 + 7 = 21$

1.3 $1.3 + (-3.3) = -2.0$ kPa.

   The resultant is negative so water is pulled into the capillary.

1.4 $290 - 25 - 125 - 80 = 60$ g

1.5 $100 - 16 - 4 - 78 = 2\%$

1.6 $\frac{4}{3} \times 3.142 \times 34 \times 34 \times 34 = 164657.56\ \mu m^3$ (2dp)

   $\frac{164657.56}{10^9} = 0.000165\ mm^3 = 1.65 \times 10^{-4}\ mm^3$

1.7 $\frac{16}{20} = 0.8$

   There are no units because this is a ratio. The worm is respiring lipid rather than glucose, which gives an RQ = 1.0.

1.8 a) $2 + 11$ ...... $1 + 10 = 11$ 13

   b) $3.2 + 8.9$ ...... $3 + 9 = 12$ 12.1

   c) $21 + 93$ ...... $20 + 90 = 110$ 114

   d) $179 + 785$ ...... $200 + 800 = 1000$ 964

   e) $2698 + 7859$ ...... $3000 + 8000 = 11\,000$ 10 557

1.9 a) $19 - 2$ ...... $20 - 1 = 19$ 17

   b) $8.2 - 0.9$ ...... $8 - 1 = 7$ 7.3

   c) $93 - 39$ ...... $90 - 40 = 50$ 54

   d) $880 - 213$ ...... $900 - 200 = 700$ 667

   e) $6789 - 1234$ ...... $7000 - 1000 = 6000$ 5555

1.10 a) 9×8 ...... $10 \times 10 = 100$ 72

   b) 5.8×5.1 ...... $6 \times 5 = 30$ 29.6

   c) 23×67 ...... $20 \times 70 = 1400$ 1541

   d) 320×190 ...... $300 \times 200 = 60\,000$ 60 800

   e) $31\,975 \times 219\,654$ ...... $(3 \times 10^4) \times (2 \times 10^5) = 6 \times 10^9$ ...... 7.02×10^9

1.11 a) $39 \div 4$ ...... $40 \div 4 = 10$ 9.75

   b) $8.9 \div 2.9$ ...... $9 \div 3 = 3$ 3.07

   c) $119 \div 6.1$ ...... $120 \div 6 = 20$ 19.5

   d) $920 \div 31$ ...... $900 \div 30 = 30$ 29.7

   e) $1\,234\,567 \div 11\,712$ ......
   $(1 \times 10^6) \div (1 \times 10^4) = 10^2 = 100$ 105.4

1.12 a) $(68 + 70 + 70 + 69 + 72) \div 5 = 69.8$

   b) His mean pulse rate at the weekend is $(64 + 62) \div 2$ $= 63$. This is lower than his weekday mean so he is more relaxed at weekends.

This makes sense. It is always worth considering if your answer is sensible. If he were more relaxed on a workday, you might infer that either his work is really boring or that he was about to do something unusually exciting that particular weekend.

1.13 $3 + 9 - (3^3 - 15) \div 2 \times 3$

   P $= 3 + 9 - (27 - 15) \div 2 \times 3$
   B $= 3 + 9 - \quad\quad 12 \quad\quad \div 2 \times 3$
   M $= 3 + 9 - \quad\quad 12 \quad\quad \div \quad 6$
   D $= 3 + 9 - \quad\quad\quad\quad 2$
   A $= \quad\quad 12 - 2$
   S $= \quad\quad 10$

1.14 $4^3 = 64$

1.15 3 genes each with 2 alleles gives $2^3 = 8$ combinations. 2 genes each with 2 alleles gives $2^2 = 4$ combinations, as above. Each of those 4 could be associated with either C or c to give ABC, ABc, AbC, Abc, aBC, aBc, abC, abc.

1.16 Three 1 in 10 serial dilutions produces a dilution of 10 $\times 10 \times 10 = 1000 = 10^3$, i.e. the suspension is 10^{-3} of the initial concentration.
   1 cm^3 diluted suspension has 250 colonies, i.e. 250 bacteria.
   1 cm^3 undiluted suspension would give 250×10^3 colonies
   so it has 250×10^3 bacteria cm^{-3} = 2.50×10^5 cm^{-3}

1.17 Length of 1st femur = 6 mm.

   Length of 3rd femur = 12 mm

   Ratio 1st femur : 3rd femur = 6 : 12 = 1 : 2.

   Long 3rd femur allows locust to jump high or far.

1.18 a) $\frac{4}{5}$ b) $\frac{2}{6} = \frac{1}{3}$ c) $\frac{4}{8} = \frac{1}{2}$

1.19 a) 782   b) 56 475   c) 2   d) 1.2478

1.20 a) 7.82   b) 0.056475   c) 0.0002   d) 12 478

1.21 3.14; 13.90; 65.21; 4.90; 8.97

1.22 $\Psi c = \psi p + \psi s = 500 + (-1500) = -1000$ kPa

1.23 $\Psi_c = \psi_p + \psi_s$ so $\Psi_p = \psi_c - \psi_s = -2000 - (-1500)$
   $= -2000 + 1500 = -500$ kPa

1.24 Ψ_c at equilibrium $= \frac{-2000 + (-1200)}{2} = \frac{-2000 - 1200}{2} = \frac{-3200}{2}$
   $= -1600$ kPa

2 Processed Numbers

2.1 % success = $\frac{\text{number produced}}{\text{total number}} \times 100 = \frac{2}{251} \times 100 = 0.797\%$ (3dp)

2.2 % mass loss = $\frac{\text{actual mass loss}}{\text{initial mass}} \times 100 = \frac{49 - 20}{49} \times 100$
$= \frac{29}{49} \times 100 = 59.2\%$

2.3 Initial count = 7 500 000 / mm^3
Subsequent count = 5 100 000 / mm^3
Change = $-$ (7 500 000 $-$ 5 100 000) = $-$ 2 400 000
% change = $\frac{\text{actual change}}{\text{initial count}} \times 100 = -\frac{2\,400\,000}{7\,500\,000} \times 100$
$= -\frac{24}{75} \times 100 = -32.0\%$

2.4

Length of each side	3
Area of each face	3 × 3 = 9
Number of faces	6
Total area	6 × 9 = 54
Volume	3 × 3 × 3 = 27
Surface area : volume ratio	54 : 27 = 2 : 1

2.5 a) For length = 2.5, sa : vol = 2.4 : 1
b) Surface area : volume = 2.5 × 2.5 × 6 : 2.5 × 2.5 × 2.5
= 37.5 : 15.625 = 2.4 : 1

2.6 a)

diameter = 2 mm	diameter = 4 mm
$\therefore$ radius = 1 mm	$\therefore$ radius = 2 mm
sa : vol = $\frac{2}{r}$: 1 = $\frac{2}{1}$: 1	sa : vol = $\frac{2}{r}$: 1 = $\frac{2}{2}$: 1
= 2 : 1	= 1 : 1

b) The worm with diameter 2 mm has a larger sa : vol ratio so for each unit of volume, it can take in more oxygen than the 4 mm worm. It can do more aerobic respiration and generate more energy. It is, therefore, likely to be more active than the 4 mm worm.

2.7 a) 2
b) 2 molecules out of a total of 32. 30 have only new strands so ratio is 1 : 15.

2.8 a) 1% = 1 g / 100 cm^3
b) 1% = 1 g / 100 cm^3
100 cm^3 has 1 g
$\therefore$ 1 cm^3 has $\frac{1}{100}$ g
$\therefore$ 10 cm^3 has $\frac{10}{100}$ g = 0.1 g
c) 1% = 1 g / 100 cm^3
$\therefore$ 0.1% solution has 0.1 g / 100 cm^3
100 cm^3 has 0.1 g
$\therefore$ 1 cm^3 has $\frac{0.1}{100}$ g
$\therefore$ 50 cm^3 has $\frac{50 \times 0.1}{100}$ g = 0.05 g
d) 1% = 1 g / 100 cm^3
$\therefore$ 2% solution has 2 g / 100 cm^3

100 cm^3 has 2 g
$\therefore$ 1 cm^3 has $\frac{2}{100}$ g
$\therefore$ 200 cm^3 has $\frac{200 \times 2}{100}$ g = 4 g

2.9 a) $CO(NH)_2$ = 12 + 16 + 2(14 + 1) = 58
58 g = 1 mole
b) C + H + H + H + H = 12 + 1 + 1 + 1 + 1 = 16
One mole CH_4 has 16 g
32 g methane has 32 ÷ 16 = 2 moles
c) C + O + O = 12 + 16 + 16 = 44
One mole has 44 g $\therefore$ 2 moles have 44 × 2 = 88 g

2.10 180 g

2.11 M_r = 180
$\therefore$ 1 dm^3 of a 1 mol dm^{-3} solution has 180 g
$\therefore$ 1 dm^3 of a 1 mmol dm^{-3} solution has $\frac{180}{1000}$ = 0.18 g

2.12 a) 1 dm^3 1 vol hydrogen peroxide produces 1 dm^3 oxygen
$\therefore$ 1 dm^3 2 vol hydrogen peroxide produces
1 × 2 = 2 dm^3 oxygen
b) 1 dm^3 1 vol hydrogen peroxide produces 1 dm^3 oxygen
$\therefore$ 1 dm^3 15 vol hydrogen peroxide produces
1 × 15 = 15 dm^3 oxygen
c) 1 dm^3 1 vol hydrogen peroxide produces 1 dm^3 oxygen
$\therefore$ 1 dm^3 15 vol hydrogen peroxide produces
1 × 15 = 15 dm^3 oxygen
$\therefore$ 4 dm^3 15 vol hydrogen peroxide produces
4 × 15 = 60 dm^3 oxygen
d) 1 dm^3 1 vol hydrogen peroxide produces 1 dm^3 oxygen
$\therefore$ 1 dm^3 5 vol hydrogen peroxide produces
1 × 5 = 5 dm^3 oxygen
$\therefore$ 1 cm^3 5 vol hydrogen peroxide produces
$\frac{5}{1000}$ = 0.005 dm^3 oxygen
$\therefore$ 20 cm^3 5 vol hydrogen peroxide produces
$\frac{20 \times 5}{1000}$ = 0.1 dm^3 oxygen

2.13 70 were counted in the 5 squares
$\therefore$ there are $\frac{70}{0.02}$ red blood cells / mm^3 in the suspension
= 3500 / mm^3

2.14 F_L = 50.0 newtons
d_L = 0.4 m
d_E = 0.05 m
$F_L \times d_L = F_E \times d_E$

$F_E = \frac{F_L \times D_L}{D_E} = \frac{50.0 \times 0.4}{0.05}$ = 400.0 newtons

3 Graphs

3.1 a) iv = temperature; dv = carbon dioxide volume

b) iv = alcohol concentration; dv = heart rate

c) iv = flow rate; dv = number of water shrimp

3.2 a) x axis: dominant hand, no scale ; y axis: number of people 0–9000

b) x axis: temperature 0–90 °C; y axis: mass 0–20g

3.3 a) continuous;   b) categorical;   c) discrete

3.4 a) 9.5%   b) 3 + 4.5 = 7.5%

3.5 a) x – pH; y– rate of casein digestion by trypsin

b) x – time at room temperature; y – bacterial count of milk

c) x – light intensity; y – area of ground ivy leaves

3.6 a) 42%   b) 4.4 kPa   c) 8.5 kPa

3.7 mass difference = mass at 10 min – mass at 5 min
= 6.6 – 3.5 = 3.1 μg

3.8 % increase = (mass at 15 min (– mass at 10 min))/
(mass at 10 min) × 100
= (9.0 – 6.6)/6.6 × 100 = 51.5%

3.9 % decrease = (length at 500 lux – length at 3000 lux)/
(length at 500 lux) × 100 = (52.5 – 14.5)/52.5 × 100
= 72.4%

3.10 Average mass of protein digested

$= \dfrac{\text{mass digested at 20 min} - \text{mass digested at 10 min}}{20 - 10} \times 100$

$= \dfrac{26.5 - 15.0}{10} \times 100 = 115\,\text{g min}^{-1}$

4 Scale

4.1 a) 1 km = 1000 m; 1 m = 1000 mm
∴ 1 km = 1000 × 1000 = 1 000 000 mm
∴ 20 km = 20 × 1 000 000 = 20 000 000
= 2×10^7 mm

b) 1 m = 1000 mm; 1 mm = 1000 μm
∴ 1 m = 1000 × 1000 = 1 000 000 μm
∴ 3.4 m = 3.4 × 1 000 000 = 3.4×10^6 μm

c) 1 m = 1000 mm; 1 mm = 1000 μm; 1 μm = 1000 nm
∴ 1 m = 1000 × 1000 × 1000 = 1 000 000 000 = 10^9 nm

d) 10^9 nm = 1 m; 1 nm = 10^{-9} m
∴ 3400 nm = 3400 × 10^{-9} = $3.4 \times 10^3 \times 10^{-9}$
= 3.4×10^{-6} m

4.2 a) 1 cm = 10 mm   ∴ 1 cm^2 = 10 × 10 = 100 mm^2

b) 1 cm = 10 mm   ∴ 1 cm^3 = 10 × 10 × 10 = 1000 mm^3

c) 1 cm = 10 mm   ∴ 1 cm^3 = 10 × 10 × 10 = 1000 mm^3
∴ 10 cm^3 = 10 × 1000 = 10 000 mm^3

d) 1 mm = 0.1 cm   ∴ 1 mm^3 = 0.1 × 0.1 × 0.1 = 0.001 cm^3
∴ 1500 mm^3 = 1500 × 0.001 = 1.5 cm^3

4.3 a) magnification = × $\frac{1}{100}$
Image length = 23 mm
Trunk length = $\frac{\text{image length}}{\text{magnification}}$

$= 23 \div \frac{1}{100}$
= 23 × 100 = 2300 mm

b) magnification = ×2
Image width = 6 mm
Body width = $\frac{\text{image width}}{\text{magnification}}$
= 6 ÷ 2 = 3 mm

4.4 magnification = × 800
Image width = 3 mm
Stoma width = $\frac{\text{image width}}{\text{magnification}}$
= 3 ÷ 800 = 0.0038 mm
As the stoma is microscopic, microns would be a more suitable unit.
Stoma width = 0.0038 × 1000 μm = 3.8 μm

4.5 a) area = 0.3 × 0.3 = 0.09 km^2

b) area = 100 × 100 = 10 000 mm^2

c) area of large face = 0.04 mm × 0.01 mm
= 40 × 10 = 400 μm^2
area of small face = 0.01 mm × 0.01 mm
= 10 × 10 = 100 μm^2
total area = (4 × 400) + (2 × 100)
= 1600 + 200 = 1800 μm^2

4.6 a) area = 500 × 500 = 250 000 mm^2 = $\frac{250\,000}{1\,000\,000}$ m^2 = 0.25 m^2

b) area = 450 × 1050 = 472 500 m^2 = $\frac{472\,500}{1\,000\,000}$
= 0.473 km^2 (3dp)

c) area = 30 × 30 = 900 mm^2 = $\frac{900}{100}$ = 9 cm^2

4.7 a) volume = 0.03 × 0.03 × 0.03 = 0.000027
= 2.7×10^{-5} mm^3
As this is a very small number, it could be turned into μm^3, by multiplying by 10^9, so that volume
= $2.7 \times 10^{-5} \times 10^9$ μm^3 = 2.7×10^4 μm^3

b) volume = 0.81 m^3, which seems an appropriate way of measuring the volume of a fish tank, so no change is needed.

5 Ratios

5.1 1 : 0 red : white

5.2 All smooth

5.3 Nn × nn where allele for normal wings = N and allele for vestigial wings = n.

5.4 3 : 1 red : white

5.5 Both heterozygous

5.6 1 : 2 : 1 red : roan : white

5.7 (AB)

5.8 (ab)

5.9 all hairy stems, red fruits

5.10 9 : 3 : 3 : 1 hairy red : hairy yellow : smooth red : smooth yellow

5.11 (AB) (aB)

5.12 (AB) (Ab) (AB) (ab)

5.13 a) Bb × bb gives offspring Bb and bb in a ratio 1 : 1.
Adding the numbers in the ratio: 1 + 1 = 2
To find how many in the '1' class in the whole litter:
$8 \div 2 = 4$

b) PpHh × PpHh gives a 9 : 3 : 3 : 1 ratio in offspring with 9 having the dominant characteristics purple and hairy.
Adding the numbers in the ratio: 9 + 3 + 3 + 1 = 16.
To find how many in the '1' class in the whole population: $480 \div 16 = 30$
To find how many in the '9' class in the whole population: $30 \times 9 = 270$

5.14 1 tall : 1 dwarf

5.15 3 Rhesus positive : 1 Rhesus negative

5.16 BbPp and bbpp where B is the allele for 'broad' and b the allele for 'narrow'; P is the allele for 'pointed' and p is the allele for 'blunt'

5.17 When inter-breeding Dexters, Dd × Dd, one quarter of the offspring will be dd and will not survive. When cross-breeding DD × Dd, there will be a ratio of 1DD : 1 Dd and all will survive.

5.18 eeAA eeAa eeaa

5.19 a) the genes are linked
b) RF = $\frac{15 + 12}{293 + 15 + 12 + 98} \times 100 = \frac{27}{418} \times 100 = 6.5\%$
The genes are 6.5 mu apart

6 The Hardy–Weinberg Equilibrium

6.1 Let the symbol for the dominant allele be B and the symbol for the recessive allele be b. 55% of alleles are B. Gene frequencies are expressed as decimals. 55% = 0.55 ∴ the allele frequency = 0.55.

6.2 $p = 0.6$
$p + q = 1$
$p = 1 - q$
$= 1 - 0.6 = 0.4$

6.3 $p = 0.6$
$f(AA) = q^2 = 0.6 \times 0.6 = 0.36$

6.4 $p = 0.8$
∴ $q = 1 - p = 1 - 0.8 = 0.2$
$f(dd) = q^2 = 0.2 \times 0.2 = 0.04$

6.5 $p = 0.61$ and $q = 0.39$
The frequency of carriers is $2pq = 2 \times 0.61 \times 0.39 = 0.48$

6.6 $p = 0.3$
∴ $q = 1 - p = 1 - 0.3 = 0.7$
$f(Rr) = 2pq = 2 \times 0.3 \times 0.7 = 0.42$

7 Statistics

7.1 i) raw   ii) processed   iii) raw
iv) processed   v) raw   vi) processed

7.2 i) 1   ii) 0   iii) 0.5   iv) 0.25

7.3 i) 4 mm / second   ii) 3.6 mm / second
iii) 60 – 79 mm²

7.4 i) mean = 34.8; median = 37
ii) mean = 1.02 mmol dm⁻³; median = 0.97 mmol dm⁻³
iii) mean = 16; median = 17

7.5 i) There is no significant correlation between the water flow rate and the number of blood worms in the river.
ii) There is no significant difference between the mean heights of soft rush growing at low or at high light intensity.
iii) The inheritance of free or attached ear lobes is Mendelian and so there is no significant difference between the observed and expected numbers of people with free or attached ear lobes.

7.6 i) yes   ii) yes   iii) no

7.7 i) $2 - 1 = 1$ ii) $(45 - 1) + (45 - 1) = 88$
iii) four phenotypic classes (red eyes + long body, white eyes + short body, red eyes + short body and white eyes + long body). Df = n − 1 = 4 − 1 = 3

7.8 i)

Number	3	6	8	9	12	15	17
Rank	1	2	3	4	5	6	7

ii)

Number	3	3	6	8	9	12	12
Rank	1.5	1.5	3	4	5	6.5	6.5

7.9 $r_{s\,calc} < r_{s\,crit}$ so the null hypothesis is accepted at the 0.05 level of significance. There is no significant correlation between light intensity and the length of the fourth internode of bramble.

7.10 The lower value of U, $U_1 = 20$ is higher than the critical value, $U_{crit} = 13$ so the null hypothesis is accepted at the 0.05 level of significance and there is no significant difference between the median numbers of water shrimp numbers on a sandy or a stony substrate.

7.11 $t_{calc} = 1.375$ and $t_{crit} = 2.101$ so $t_{calc} < t_{crit}$. The null hypothesis is accepted at the 0.05 level of significance and so there is no significant difference between the mean numbers of mayfly nymphs in high or low concentrations of dissolved oxygen.

7.12 $\chi^2_{crit} = 3.84$ and $\chi^2_{calc} = 4.68$ so $\chi^2_{crit} < \chi^2_{calc}$. The null hypothesis is rejected at the 0.05 level of significance and inheritance is not Mendelian.

Test yourself answers

1 Numbers

1
$$T = 34\%$$
$$A = T$$
$$\therefore \quad A = 34\%$$
$$A + T = 100 - (G + C)$$
$$\therefore \quad A + T = 34 + 34 = 68\%$$
$$\therefore \quad G + C = 100 - 68 = 32\%$$
$$G = C$$
$$G = 32/2 = 16\%$$

2 Difference = maximum expiration + maximum inspiration
$$= 0.29 + 0.30 = 0.59 \text{ kPa}$$

3 One cardiac cycle takes $(1.34 - 0.50) = 0.84$ s
In 0.84 s is one cardiac cycle
∴ in 1 s is 1/0.84 cycles
∴ in 60 s are $60 \times 1/0.84 = 71.4$
∴ heart rate is 71.4 bpm

4 Blood flow during exercise : blood flow at rest
$$= 1200 : 300$$
$$= 4 : 1$$

5 a) Rate of infection in unsprayed plants =
$(15.5 - 8.5) / 5 = 1.4$ AU / day
b) Rate of infection in sprayed plants =
$(7.7 - 7.0) / 5 = 0.14$ AU / day
Ratio of infection rate of sprayed : unsprayed plants
$0.14 : 1.4 = 1 : 10$

6 Rate $= \dfrac{0.78 - 0.06}{4 - 1} = \dfrac{0.72}{3} = 0.24$ AU / day

7 a) There is uptake both with and without sodium.
b) Uptake is greater with sodium than without sodium.
c) Between 5 and 45 minutes, without sodium, uptake increases by a factor of $(33/3) = 11$.
d) Between 5 and 45 minutes, with sodium, uptake increases by a factor of $(630/70) = 9$.

8 $109.6 - 96.0 = 13.6$ kg/ha

9 Mean number per 0.1×0.1 m square = (3 + 4 + 2 + 2 + 0 + 2 + 4 + 6 + 3 + 1 + 2 + 5 + 2 + 0 + 2 + 1+ 3 + 3 + 0 + 0 + 2 + 4 + 7 + 5 + 3) ÷ 25 = 2.6

There are on average 2.6 wood sages per 0.1×0.1 $= 0.01$ m^2
∴ in 1 m^2 there are $2.6 \div 0.01 = 260$ plants per m^2
Density = 260 plants per m^2

10 There are 18 bases in this sequence
Three bases code for one amino acid
∴ This sequence codes for $18/3 = 6$ amino acids

11 $G = C$ ∴ cow DNA has 21% guanine
$G = C$ ∴ yeast DNA has 19% guanine
$G + C = 19 + 19 = 38\%$
∴ $A + T = 100 - 38 = 62\%$
$A = T = 62/2 = 31\%$

Organism	Percentage of each base in DNA			
	thymine	cytosine	adenine	guanine
cow	29	21	29	21
yeast	31	19	31	19

12 Growth rate $= \dfrac{\text{number on day 12} - \text{number on day 0}}{12 - 0}$
$$= \dfrac{310 - 30}{12} = \dfrac{280}{12} = 23.3 \text{ plants/day}$$

13 Total amino acid production for first 3 hours =
$95 \times 3 = 285$ mg dm^{-3}
Total amino acid = total production for first 3 hours + production in next 5 hours
$= (285 + 246) = 531$ mg dm^{-3}
Mean rate of amino acid production over the whole 8 hours = $531/8 = 66.4$ mg dm^{-3} h^{-1}

14 $38 \times 30.7 = 1166.6$ kJ mol^{-1}

2 Processed Numbers

1

Substance	Passing through proximal coiled tubule	Passing through collecting duct	Quantity reabsorbed	% reabsorbed
Water	206 dm^3	1.7 dm^3	204.3 dm^3	$\frac{\text{actual volume reabsorbed}}{\text{volume through PCT}} \times 100$ $= \frac{204.3}{206} \times 100$ $= 99.17\%$
Urea	206 g	206 – 206 = 0 g	206 g	$\frac{\text{actual volume reabsorbed}}{\text{volume through PCT}} \times 100$ $= \frac{206}{206} \times 100 = 100\%$
Glucose	61 g	29 g	61 – 29 = 32 g	$\frac{\text{actual volume reabsorbed}}{\text{volume through PCT}} \times 100$ $= \frac{32}{61} \times 100 = 52.5\%$

2 100 cm^3 blood contains 20 cm^3 oxygen
70% of this is lost, i.e. 70% of 20 = $\frac{70}{100} \times 20 = 14$ cm^3
∴ remaining oxygen = (20 – 14) = 6 cm^3/100 cm^3 blood

3 Initial mass = 58 g
Mass lost = 6% of 58 g = $\frac{6}{100} \times 58 = 3.48$ g
∴ final mass = 58 – 3.48 = 54.52 g

4

2 mm cube	5 mm cube
Area of face = 2 × 2 = 4 mm^2	Area of face = 5 × 5 = 25 mm^2
Total area = 6 × 4 = 24 mm^2	Total area = 6 × 25 = 150 mm^2
Volume = 2 × 2 × 2 = 8 mm^3	Volume = 5 × 5 × 5 = 125 mm^3
Surface area : volume ratio = 24 : 8 = 3 · 1	Surface area : volume ratio = 150 : 125 – 6 : 5 = 1.2 : 1

2 mm cube has a larger surface area : volume ratio. So each unit of volume in the 2 mm cube will take in more water than in the 5 mm cube ∴ the % uptake by the 2 mm cube is greater.

5 Add 30 cm^3 distilled water.

6 10 cm^3 blood made 18 m^2 phospholipids in a monolayer
∴ 1 cm^3 blood made $\frac{18}{10} = 1.8$ m^2 in a monolayer
∴ 1 mm^3 blood made 1.8×10^{-3} m^2 in a monolayer
7.5×10^6 red blood cells made 1.8×10^{-3} m^2 in a monolayer

∴ 1 cell made $(1.8 \times 10^{-3})/(7.5 \times 10^6)$ in a monolayer
$= \frac{1.8}{2 \times 7.5} \times 10^{-9}$ m^2 in a bilayer
$= 0.12 \times 10^{-9}$ m^2 in a bilayer
1 m = 1 000 000 μm ∴ 1 m^2 = 10^{12} μm^2
0.12×10^{-9} m^2 = $0.12 \times 10^{-9} \times 10^{12}$ μm^2 = 0.12×10^3
= 120 μm^2 in a bilayer from one cell
∴ area of one red blood cell = 120 μm^2

7 Britain: percentage of bat species = $\frac{18}{1240} \times 100 = 1.5\%$ (1dp)
Indonesia: percentage of bat species = $\frac{175}{1240} \times 100$ = 14.1% (1dp)
Possible reasons: in Indonesia there may be more niches; fewer predators; less habitat destruction; fewer environmental toxins

8 1 cm^3 of 10^{-4} dilution produced 72 colonies
∴ 1 cm^3 of undiluted sample would produce 72×10^4 colonies
∴ the bacterial concentration in the original ice cream was 7.2×10^5 bacteria cm^{-3}

NOTICE THE USE OF STANDARD NOTATION, WITH ONE DIGIT BEFORE THE DECIMAL POINT

9 A 1000 dilution was made and 100 were counted in the 5 squares
∴ there are $\frac{100}{0.02}$ bacteria / mm^3 in the 10^{-3} dilution
∴ there are $\frac{100}{0.02} \times 1000 = 5\,000\,000 = 5 \times 10^6$ / mm^3 in the undiluted suspension.

3 Graphs

1 Continuous variation – sample is approximately normally distributed.

2 a)

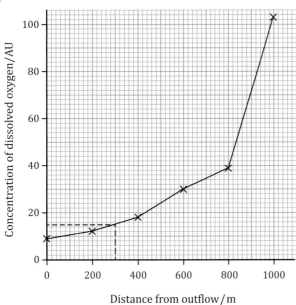

b) From the graph, at 300 m
oxygen concentration = 15 AU

3

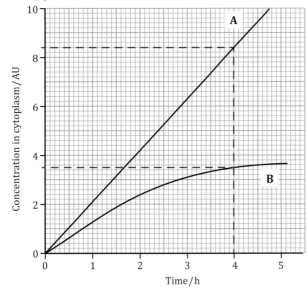

Drawing on the graph shows:
Uptake at A at 4 h = 8.4 AU
Uptake of B at 4 h = 3.5 AU
Difference = 8.4 – 3.5 = 4.9 AU
You can give the answer to 1 decimal place as the graph
allows you to read to this level of accuracy.

4

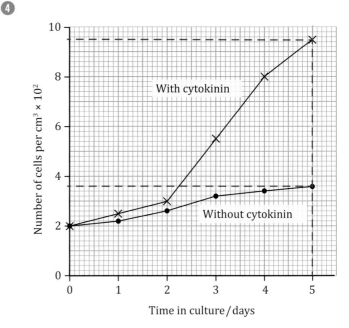

a) % increase at 5 days

$$= \frac{\text{cell number at 5 days – cell number at 0 days}}{\text{cell number at 0 days}} \times 100\%$$

$$= \frac{950 - 200}{200} \times 100\% = 375\%$$

b) % increase at 5 days due to cytokinin =

$$\frac{\text{cell number at 5 days with cytokinin – cell number at 5 days without cytokinin}}{\text{cell number at 5 days without cytokinin at 0 days}} \times 100\%$$

$$= \frac{950 - 360}{360} \times 100\% = 164\% \text{ (0 dp)}$$

5

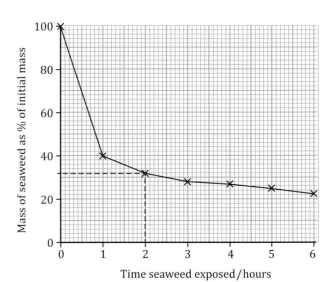

% lost = 100 – 32 = 68%
Actual mass lost = 68% of 250 g
$$= \frac{68}{100} \times 250 = 170 \text{ g}$$

6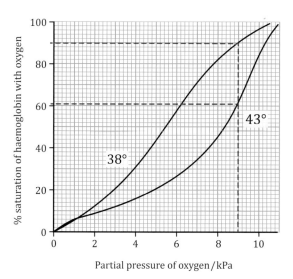

a) from the graph, blood at 38°C with haemoglobin that is 90% saturated would be 61% saturated at 43°C

∴ it would lose 90 – 61 = 29% of its oxygen

b) blood with 100% saturated haemoglobin contains 105 cm³ oxygen

∴ 1% saturated haemoglobin contains 105/100 = 1.05 cm³ oxygen

∴ 90% saturated haemoglobin contains 1.05 × 90 = 94.5 cm³ oxygen

From a), the haemoglobin loses 29% of its oxygen

∴ it loses $\frac{29}{100}$ × 94.5 = 27.4 cm³ oxygen

7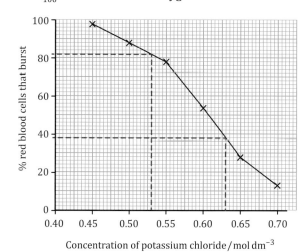

% difference = % burst at 63 mol dm⁻³ KCl

                         – % burst at 53 mol dm⁻³ KCl

= 82 – 38 = 44%

4 Scale

1 Using the standard three statements:

At a ×4 objective lens

1   6 epu = 20 smu

2   1 smu = 0.01 mm

3   1 epu = 20/6 smu

           = 20/6 × 0.01 mm

           = 3.33 × 0.01 mm

           = 0.033 mm

Very small numbers of mm are better expressed as micrometres (μ or μm) so you can complete the calculation like this:

           = 0.033 × 1000 μm

           = 33 μm

2 Using the standard three statements:

At a ×4 objective lens

1   45 epu = 40 smu

2   1 smu = 0.01 mm

3   1 epu = 40/45 smu

           = 40/45 × 0.01 mm

           = 0.89 × 0.01 mm

           = 0.0089 mm

Very small numbers of mm are better expressed as micrometres (μ or μm) so you can complete the calculation like this:

           = 0.0089 × 1000 μm

           = 8.9 μm

3 diameter = 20 μm

∴   radius = $\frac{20}{2}$ = 10 μm

    volume = $\frac{4}{3}\pi r^3$

           = $\frac{4}{3}$ × 3.142 × 10 × 10 × 10 μm³

           = 4189.33 μm³ (2dp)

4 Width measured on diagram = 20 mm

Magnification = 3 × 10⁶

Actual width = $\frac{\text{measured width}}{\text{magnification}}$ = $\frac{20}{3 \times 1000\,000}$ × 1000 μm

           = 0.007 μm (3dp) = 0.007 × 1000 nm

           = 7 nm

5 Measured maximum length of vacuole = 24 mm

= 24 × 1000 μm

Magnification = $\frac{\text{image length}}{\text{object length}}$ = $\frac{24 \times 1000}{40}$ = 600

6 Magnification = $\frac{\text{image length}}{\text{object length}}$

Nucleus diameter = 11 mm = 11 × 1000 μm

Nucleus diameter = $\frac{\text{image diameter}}{\text{magnification}}$ = $\frac{11 \times 1000}{1000}$ μm = 11 μm

5: Ratios

1 a) 3 chocolate : 1 white

b) Cc and Cc

c)

	Ⓒ	ⓒ
Ⓒ	CC chocolate	Cc chocolate
ⓒ	Cc chocolate	cc white

d) 1

2

Parents	green GG	yellow gg
Gametes	(G)	(g)
F₁		Gg Green

Test cross:	Gg green F1	gg yellow
Gametes	(G)(g)	(g)

F₂

	(g)
(G)	Gg green
(g)	gg yellow

1:1 green:yellow

3 Showing the information in the question as a pedigree or family tree:

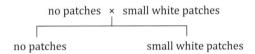

The question states that cats with no patches have the genotype S^1S^1, so the genotypes of one parent and one of the offspring are known. The kitten with small white patches must have inherited the S^1 allele from its parent with no patches. It cannot have inherited the same allele from its other parent because then it would be S^1S^1 and have no patches. It must therefore have inherited the S^2 allele from the other parent. So cats with small white patches have the genotype S^1S^2.

Parents	no patches S^1S^1	small white patches S^1S^2
Gametes	(S¹)	(S¹)(S²)

F₁

	(S¹)
(S¹)	S^1S^1 no patches
(S²)	S^1S^2 small white patches

4 The allele for mauve flowers is M; the allele for white flowers is m.

The allele for prickly fruit is P; the allele for smooth fruit is p.

The $9:3:3:1$ ratio in the F_2 is produced by crossing two F_1 individuals that are heterozygous at two genes. The parents were therefore homozygous dominant or homozygous recessive at both genes.

Parents	mauve, prickly MMPP	white, smooth mmpp
Gametes	(MP)	(mp)
F₁		MmPp mauve, prickly

F₁ cross:	MmPp mauve, prickly	MmPp mauve, prickly
Gametes	(MP)(Mp)(mP)(mp)	(MP)(Mp)(mP)(mp)

F₂

	(MP)	(Mp)	(mP)	(mp)
(MP)	MMPP mauve, prickly	MMPp mauve, prickly	MmPP mauve, prickly	MmPp mauve, prickly
(Mp)	MMPp mauve, prickly	MMpp mauve, smooth	MmPp mauve, prickly	Mmpp mauve, smooth
(mP)	MmPP mauve, prickly	MmPp mauve, prickly	mmPP white, prickly	mmPp white, prickly
(mp)	MmPp mauve, prickly	Mmpp mauve, smooth	mmPp white, prickly	mmpp white, smooth

Phenotypic ratio 9 mauve prickly : 3 mauve smooth : 3 white prickly : 1 white smooth

5 a)

	(rs)
(RS)	RrSs
(Rs)	Rrss
(rS)	rrSs
(rs)	rrss

b) There would be more parental ((RS) and (rs)) gametes; there would be some recombinant gametes ((rS) and (Rs)); because of crossing over between the genes; the four gametes would not be in equal proportion; so the ratio will not be $1:1:1:1$ as expected with unlinked genes.

6 $RF = \dfrac{23 + 21}{23 + 21 + 165 + 191} \times 100 = \dfrac{44}{400} \times 100 = 11\%$

7 a) 9 black rough : 3 black smooth : 3 white rough : 1 white smooth

b) the genes for fur colour and texture are linked

c) cross-over value $= \dfrac{6 + 4}{58 + 6 + 4 + 20} \times 100 = \dfrac{10}{88} \times 100 = 11.4\%$

The two genes are 11.4 map units apart.

8

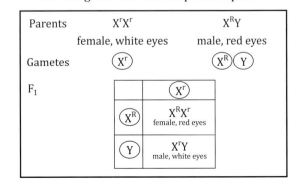

Parents	X^rX^r female, white eyes	X^RY male, red eyes
Gametes	(X^r)	(X^R)(Y)

F₁

	(X^r)
(X^R)	X^RX^r female, red eyes
(Y)	X^rY male, white eyes

(9)

K   M   N   L
| | | |
  15   10   8

6 The Hardy-Weinberg Equilibrium

(1)

a) 0.2% are non-tasters $0.2 \times \frac{1}{100} = 0.002$ are non-tasters.

b) 0.002 of the population are non-tasters and are homozygous recessive

$\therefore q^2 = 0.002$

$\therefore q = \sqrt{0.002} = 0.045$ (3dp)

c) $p + q = 1$

$\therefore p = 1 - q = 1 - 0.045 = 0.955$

d) frequency of heterozygotes $= 2pq$

$= 2 \times 0.955 \times 0.045 = 0.086$

e) frequency of tasters $= p^2 + 2pq$

$= 0.955^2 + (2 \times 0.955 \times 0.045)$

$= 0.912 + 0.086 = 0.998$

(2) Frequency of affected individuals $= \frac{1}{200} = 0.005$

If homozygous recessive allele produces the condition,

$q^2 = 0.005$

$\therefore \quad q = 0.07$ (2 dp)

$p + q = 1$

$\therefore \quad p = 1 - 0.07 = 0.93$

Frequency of carriers $= 2pq = 2 \times 0.93 \times 0.07 = 0.13$

(3) 58% of people are MM $= 0.58$ of the population

$p^2 = 0.58$

$\therefore p = \sqrt{0.58} = 0.76 =$ frequency of M

36% of people are MN $= 0.36$ of the population

$2pq = 0.36$

$p = 0.76 \therefore q = \frac{0.36}{2 \times 0.76} = 0.24 =$ frequency of N

As a check:

100 – (58 + 36) of people are NN $= 100 - 94 = 6\%$ of people $= 0.06$ of the population

$q^2 = 0.06 \therefore q = \sqrt{0.06} = 0.24$

As another check:

$p = 0.76$

$p + q = 1$

$\therefore q = 1 - p = 1 - 0.76 = 0.24$

(4) 64% are rollers $\therefore$ 64% have genotypes RR or Rr

$p^2 + 2pq = 0.64$

$\therefore q^2 = 1 - 0.64 = 0.36$

$\therefore q = \sqrt{0.36} = 0.6$

$p + q = 1$

$\therefore p = 1 - q = 1 - 0.6 = 0.4$

$f(rr) = q^2 = 0.36 \therefore rr = 36\%$

$f(RR) = p^2 = 0.16 \therefore RR = 16\%$

$f(Rr) = 100 - (36 + 16) = 100 - 52 = 48\%$

As a check:

$F(Rr) = 2pq = 2 \times 0.4 \times 0.6 = 0.48 \therefore Rr = 48\%$

7 Statistics

(1)

a) A correlation is being sought; there are ten pairs of readings

b) There is no significant correlation between soil pH and the % area cover of heather (*Erica tetralix*)

c)

Soil pH	Rank$_1$	% area cover of heather	Rank$_2$	Difference between ranks $R_1 - R_2 = d$	d^2
4.5	1	100	10	–9	81
5.0	2	97	9	–7	49
5.5	3	92	7.5	–4.5	20.25
6.0	4	92	7.5	–3.5	12.25
6.5	5	76	5	0	0
7.0	6	83	6	0	0
7.5	7	70	4	3	9
8.0	8	61	1	7	49
8.5	9	66	3	6	36
9.0	10	64	2	8	64
				$\Sigma d = 0$	320.5

$r_s = 1 - \frac{6\Sigma d^2}{n(n^2 - 1)} = 1 - \frac{6 \times 320.5}{10 \times 99} = 1 - \frac{1923}{990} = 1 - 1.942$

$= -0.942$

r_s is negative $\therefore$ the correlation is negative.

For $n = 10$ at 0.05 level of significance,

$r_{s\,crit} = 0.6485$.

$r_{s\,calc} > r_{s\,crit} \therefore$ the null hypothesis is rejected at the 0.05 level of significance.

As the soil pH increases above 4.5, the % area cover of *Erica tetralix* decreases.

(2) Null hypothesis: there is no significant correlation between the diameter of the callus of tobacco cells and the time for which it has been in culture.

x	x^2	y	y^2	xy
0	0	2	4	0
2	4	2	4	4
4	16	3	9	12
6	36	14	196	84
8	64	6	36	48
10	100	8	64	80
12	144	12	144	144
14	196	13	169	182
16	256	15	225	240
18	324	16	256	288
$\Sigma x = 90$	$\Sigma x^2 = 1140$	$\Sigma y = 91$	$\Sigma y^2 = 1107$	$\Sigma xy = 1082$

Substituting into the equation:

$$r = \frac{n(\Sigma xy) - (\Sigma x)(\Sigma y)}{\sqrt{\{[n\Sigma x^2 - (\Sigma x)^2][n\Sigma y^2 - (\Sigma y)^2]\}}}$$

where r = product-moment correlation coefficient and n = number of samples.

$$r = \frac{10(1082) - (90)(91)}{\sqrt{\{[10 \times 1140 - (90)^2][10 \times 1107 - (91)^2]\}}}$$

$$= \frac{10\,820 - 8190}{\sqrt{\{[11\,400 - 8100][11\,070 - 8281]\}}}$$

$$= \frac{2630}{\sqrt{\{[3300][2789]\}}} = \frac{2630}{3034} = 0.87$$

For $n = 10$ at the 0.05 level of significance, $r_{crit} = 0.5494$. $r_{calc} > r_{crit}$ so the null hypothesis is rejected at the 0.05 level of significance and the diameter of the tobacco callus is positively correlated with its time in tissue culture.

r_{calc} is positive so the correlation is positive. The coefficient of determination, $r^2 = 0.87^2 = 0.76$ so there is a strong positive correlation.

3 a) The Mann-Whitney U test is suitable as there is a small sample of non-normally distributed data.

   b) The null hypothesis is that there is no significant difference in the median number of larvae at sites 1 and 2.

   c) The table shows the data ordered and assigned ranks:

Site 1					9	9	10		15	16	21	
Rank $_1$					6.5	6.5	8		10	11	12	$\Sigma R_1 = 54$
Site 2	2	3	5	6	8			12				
Rank $_2$	1	2	3	4	5			9				$\Sigma R_2 = 24$

Calculating values for U

$$U_1 = n_1 \times n_2 + \tfrac{1}{2} n_2(n_2 + 1) - \Sigma R_2$$
$$= 6 \times 6 + \tfrac{1}{2} \times 6(6 + 1) - 24 = 33$$
$$U_2 = n_1 \times n_2 + \tfrac{1}{2} n_1(n_1 + 1) - \Sigma R_1$$
$$= 6 \times 6 + \tfrac{1}{2} \times 6(6 + 1) - 54 = 3$$

For $n_1 = n_2 = 6$, $U_{crit} = 5$
The lower value of U is $U_2 = 3$. $U_2 < U_{crit}$ so the null hypothesis is rejected at the 0.5 level of significance and the median number of larvae in site 2 is less than the median number at site 1.

4

Number of lymphocytes	
Sample A	Sample B
176	150
189	167
165	143
157	189
187	158
196	154
168	147
176	168
165	155
196	187
Mean 177.5	161.8

Equation $t = \dfrac{\bar{x}_A - \bar{x}_B}{\sqrt{\dfrac{s_A^2}{n_A} + \dfrac{s_B^2}{n_B}}}$

$$= \frac{177.5 - 161.8}{\sqrt{\dfrac{227.36}{10} + \dfrac{173.45}{10}}}$$

$$= \frac{15.7}{\sqrt{40.081}} = \frac{15.7}{6.331} = 2.480$$

Number of degrees of freedom = $(10 - 1) + (10 - 1) = 18$
For 0.05 level of significance, $t_{crit} = 2.101$
$t_{calc} = 2.480$

$t_{calc} > t_{crit}$ so the null hypothesis is rejected at the 0.05 level of significance and so person A has more lymphocytes than person B.

5 Null hypothesis: inheritance is Mendelian, with a ratio of 3 purple stems : 1 green stem

Stem colour	Observed (O)	Expected (E)	$O - E$	$(O - E)^2$	$\dfrac{(O - E)^2}{E}$
Purple	346	$\tfrac{3}{4} \times 480 = 360$	-14	196	0.544
Green	134	$\tfrac{1}{4} \times 480 = 120$	14	196	1.633

$$\chi^2_{calc} = \sum \frac{(O - E)^2}{E} = 0.544 + 1.633 = 2.177.$$

From the table, for 1 degree of freedom, at 0.05 level of significance, the critical value of $\chi^2_{crit} = 3.841$. $\chi^2_{calc} < \chi^2_{crit}$ so the null hypothesis is accepted at the 0.05 level of significance. Inheritance is Mendelian and any deviation from the ideal ratio is due to chance.

Glossary

Accuracy the closeness of a reading to the true value.

Allele frequency the proportion of a given allele in a population.

Antilog Bar chart graph with bars representing discrete data, where the length of each bar is proportional to the values it represents.

Biodiversity a measure of the number of species and the number within each species present in a system.

BMI body mass index $= \dfrac{\text{mass} / \text{kg}}{\text{height}^2 / \text{m}^2}$.

BODMAS the order of arithmetic operations: brackets, orders, divide and multiply, add and subtract.

Calorimetry measurement of the energy content of a substance, given by the equation:
energy = mass × specific heat capacity × temperature rise.

Cell cycle repeated sequence of processes undergone by a cell from one division to the next.

CentiMorgan the separation of two genes on a chromosome that gives a recombination frequency of 1%; 1 map unit.

Chi squared (χ^2) test statistical test used to test if an observed outcome is sufficiently close to an expected outcome to consider deviation from the prediction to be due to chance, rather than a biological mechanism.

Co-dominance situation describing two alleles of a gene where neither allele is dominant and the heterozygote shows expression of both alleles, so has a different phenotype from either homozygote.

Concentration the mass of a solute in a given volume of solution. It is expressed in mol dm^{-3}, % (g/100 cm^3) or, in the case of hydrogen peroxide, vol.

Confidence interval the range of values in which the actual value is likely to occur.

Confidence limits the upper and lower values of the confidence interval.

Confidence level the probability associated with a confidence interval.

Continuous data data that does not fall into discrete classes but may take any value within a given range.

Correlation the dependence of one value on another.

Critical value a threshold value with which a value calculated from test data is compared, to determine whether a null hypothesis is accepted or rejected.

Cross-over value also called the recombination frequency. Indicates how often a cross-over event between two genes in prophase I of meiosis is represented in the phenotype of offspring. It is shown by the equation $\text{RF} = \dfrac{\text{number of recombinants}}{\text{total number of progeny}} \times 100$.

Degrees of freedom the number of values that can vary independently in a statistical calculation.

Dihybrid inheritance the simultaneous inheritance of the alleles of two genes.

Dilution a reduction in concentration, often expressed as a percentage of a stock solution.

Discontinuous data data that can take only a small number of discrete values.

Disney's index a measure of biodiversity suitable for describing plant communities.

Distribution, bimodal a distribution of values where there are two distinct modes, appearing as two separate peaks on a probability distribution curve.

Distribution, normal a distribution of values symmetrical around the mean, median and mode, which are identical.

Distribution, skewed a distribution in which more values tend towards one extreme than the other, such that the mean, median and mode are not identical.

DNA polymer of deoxyribonucleotides, comprising hereditary information in cells and some viruses.

Ecological pyramids diagrams drawn as a stack of horizontal bars representing the trophic levels in a community, where the area of each bar is proportional to the number, biomass or energy flowing through each trophic level.

Epistasis situation in which the alleles of one gene suppress or allow the expression of the alleles of another gene.

Frequency histogram graph showing the probability distribution of continuous categories or intervals of data, where the length of each bar is proportional to the frequency of that category.

Genetic code three-base sequences that direct either the addition of specific amino acids to an amino acid chain or that terminate the extension of a growing amino acid chain.

Genetic mapping the identification of the position of genes with respect to each other on a chromosome.

Gross primary productivity (GPP) rate of incorporation of energy into the products of photosynthesis, often expressed as $kJ/m^2/y$.

Hardy–Weinberg equilibrium relationship between two alleles of a given gene in a stable population. Described by the equation $p^2 + 2pq + q^2 = 1$.

Histogram chart showing frequency of consecutive classes of values.

Incomplete dominance situation describing two alleles of a gene where neither allele is dominant and the heterozygote has a different phenotype from either homozygote, which may appear to be intermediate between the two homozygote phenotypes.

Indices powers, exponents. An index is the number of times a number, the base, must be multiplied by itself.

Kite diagram graphical representation of plant distribution along a transect, shown symmetrically about a distance axis, representing the transect.

Lethal recessive recessive allele that would be lethal if homozygous.

Level of significance the probability of rejecting a null hypothesis when, in fact, it should be accepted.

Linkage relationship between genes on the same chromosome, resulting in the alleles on homologous chromosomes having non-random distribution in the gametes.

Lincoln index estimation of an animal population using counts from a mark-release-recapture experiment. The estimated population

$$= \frac{\text{number in 2nd sample} \times \text{number in 1st sample}}{\text{number marked in 2nd sample}}.$$

Magnification the number of times an image size is greater than that of an object, where

$$\text{magnification} = \frac{\text{image size}}{\text{object size}}.$$

For an image viewed in a microscope, magnification = objective magnification × eyepiece magnification.

Mann-Whitney U test statistical test to determine if the median values of two sets of data are close enough to be considered equivalent.

Map unit distance between two genes on a chromosome, measured in centiMorgans.
1 map unit = 1 cM.

Mean, arithmetic the sum of numbers divided by the number of numbers.

Median the middle value of an odd set of numbers or the arithmetic mean of the two middle values of an even set of numbers.

Mendelian ratio ratio of phenotypes produced under ideal conditions, including those in which gametes are produced in equal numbers and fuse at random, with no differential viability of gamete or zygote, as described by the results of Gregor Mendel.

Mode the value that appears most often in a set of numbers.

Molarity for a solution, the number of moles present in $1 \, dm^3$ solution, at standard temperature and pressure.

Mole the mass, in grams, of an atom's, ion's or molecule's relative molecular mass.

Monohybrid inheritance the inheritance of a characteristic determined by the alleles of one gene.

Negative number real number less than zero.

Net (or nett) primary productivity (NPP) the energy in molecules incorporated into plant biomass and available to the second trophic level. Expressed in $kJ/m^2/y$. Described by equation NPP = GPP − R, where R represents energy lost when substrates are respired.

Nomogram a drawing of scales showing how three or more variables relate to each other, allowing prediction of a value when two others are known and a line joining them on two scales is extended to the third.

Non-Mendelian ratio ratio of phenotypes that does not correspond to those described by Mendel's work, and is derived from non-standard mechanisms, including epitasis, linkage, lethal allele combinations and differential viability of gametes, zygotes or embryos.

Null hypothesis statement that there is no significant statistical relationship between two measured variables.

One-tailed test a statistical test in which a spread of data is either more or less than a given value but could not be both.

Oxygen dissociation curve graph showing the percentage saturation of haemoglobin with oxygen at different partial pressures of oxygen.

Partial dominance general term for a situation in which neither of two alleles of a gene shows dominance over the other and the heterozygote has a different phenotype from either homozygote. Refers to both incomplete dominance and to co-dominance.

Pearson linear correlation test, PMCC or PPMCC a statistical test to measure the degree of linear relationship between two variables.

Per cent area cover estimate of the per cent of an area covered by plants.

Per cent frequency the number of grid squares of a gridded quadrat in which a plant species occurs, or the number of times a pin touches a plant of a given species in a point frame quadrat, expressed as a percentage.

Pie chart a circular chart in which the area of each sector is proportional to the quantity of that category.

Probability likelihood of an event.

Punnett square table showing frequency of possible allele combinations following fertilisation, assuming all gametes are produced at same frequency and fuse at random.

Pyramid of biomass diagram drawn as a stack of horizontal bars representing the trophic levels in a community, where the area of each bar is proportional to the fresh or dry biomass at that trophic level.

Pyramid of energy diagram drawn as a stack of horizontal bars representing the trophic levels in a community, where the area of each bar is proportional to the energy flowing through that trophic level.

Pyramid of numbers diagram drawn as a stack of horizontal bars representing the trophic levels in a community, where the area of each bar is proportional to the number of individuals at that trophic level.

Range difference between the lowest and highest value of a data set.

Ranking placing data in numerical order and assigning a value to their position in that order.

Ratio a relationship between two numbers of the same kind.

Recombination frequency, cross-over value indicates how often a cross-over event between two genes in prophase I of meiosis is represented in the phenotype of offspring. It is shown by the equation

$$RF = \frac{\text{number of recombinants}}{\text{total number of progeny}} \times 100.$$

Reliability, repeatability, consistency the closeness of replicate readings to each other.

Replicability the likelihood of making identical readings if an experiment is carried out in exactly the same way with exactly the same equipment.

Respiratory quotient (RQ) the ratio of volume of carbon dioxide evolved and oxygen absorbed, given by the equation $RQ = \dfrac{\text{volume of carbon dioxide evolved}}{\text{volume of oxygen absorbed}}$

Rounding replacing a number with a shorter or simpler approximate value that may indicate the level of its accuracy.

Sampling selection of a set of individuals that is taken to represent the whole population.

Scale bar a line associated with an image to indicate the actual length of the same length on the image.

Scale break a line drawn on the axis of a graph showing a discontinuity in the scale, separating higher from lower values.

Serial dilution process of diluting a solution or suspension step-wise, so that one dilution is used to make the successive dilution.

Sex linkage expression of an allele predominately in either males or females, dependent on the gene being on a sex chromosome.

Significant figures the digits in a number that make a contribution to its precision

Simpson's index a measure of biodiversity suitable for describing animal communities.

Spearman rank correlation test statistical test to determine if two sets of ranked data are dependent upon each other.

Species evenness a diversity index that describes how close the numbers of each species present in a habitat are.

Species richness the number of species in a habitat.

Spirometer apparatus to measure the volume of air inhaled or exhaled.

Standard deviation a measure of the dispersion of data around the mean.

Standard error of mean standard deviation of all the calculated values of the mean.

Standard notation, standard form expression of a number using a 'mantissa' from 1 up to, but not including, 10 multiplied by 10 with a power, either positive or, for numbers smaller than 1, negative.

t **test** statistical test to determine if the arithmetic means of two sets of data are close enough to be considered equivalent.

Tangent a line that shows the gradient of a curve at the point of contact; it is parallel to the curve at that point.

Test cross cross between an individual of unknown genotype and an individual carrying only recessive alleles for the gene or genes under consideration.

Tidal volume volume of air inhaled or exhaled during breathing while the body is at rest.

Two-tailed test a statistical test in which the data can be either more or less than a given value.

Variation the phenotypic differences between members of the same species.

Variance a measure of how widely spread the values of a variable are.

Vital capacity volume of air exhaled by forced breathing out.

Specification map

This table shows the mathematical requirements of various examination boards for the post-16 Biology specifications, taught from 2015.

The absence of a tick does not imply the skill is not required, rather that it has not been explicitly stated. It is important to read the relevant specification, which provides more details and presents the context of each statement.

Statements in bold are tested in the second year of an A level course, but not in the first year.

Topic	Section	AQA	CCEA	Eduqas	Edexcel	OCR	WJEC	Cambridge pre-U	Edexcel International	IB
Recognise and use appropriate units in calculations	2.3 4.1 4.2	✓		✓	✓	✓	✓			
Recognise and use expressions in decimal and standard form	1.1 1.4 1.6 1.7	✓	✓	✓	✓	✓	✓	✓	✓	✓
Use ratios, fractions and percentages	1.5 2.1 3.6.2 3.7	✓	✓	✓	✓	✓	✓	✓	✓	✓
Use scales for measuring	2.1.5	✓	✓	✓	✓	✓	✓		✓	✓
Estimate results to check that calculated values are appropriate	1.1.5	✓	✓	✓	✓	✓	✓	✓	✓	✓
Use calculators to find and use exponential and logarithmic functions	1.4 3.6.4 including numbers of chromosome combinations after meiosis & fertilisation and pH from pH=−log10 H+	✓	✓	✓	✓	✓	✓	✓	✓	

Topic	Section	AQA	CCEA	Eduqas	Edexcel	OCR	WJEC	Cambridge pre-U	Edexcel International	IB
Handling data										
Use an appropriate number of significant figures — Report calculations to an appropriate number of significant figures given raw data with varying numbers of significant figures	1.7.4	✓	✓	✓	✓	✓	✓	✓	✓	✓
Understand that calculated results can only be reported to the limits of the least accurate measurement	2.1.5	✓	✓	✓	✓	✓	✓	✓	✓	✓
Construct and interpret frequency tables and diagrams, bar charts and histograms — Represent and interpret a range of data in a table with clear headings, units and consistent number of decimal places, and in suitable graphs	3.3 3.7 3.8	✓	✓	✓	✓	✓	✓	✓	✓	✓
Understand simple probability — Use the terms probability and chance appropriately	5.2–5.5 7.3	✓	✓	✓	✓	✓	✓	✓	✓	✓
Understand the principles of sampling as applied to scientific data — Analyse random data, e.g. as collected by Simpson's Diversity Index	2.4 7.2 7.15	using $D = \frac{N(N-1)}{\Sigma n(n-1)}$	using $D = \frac{\Sigma n(n-1)}{N(N-1)}$	using $D = 1 - \frac{\Sigma n(n-1)}{N(N-1)}$	using $D = \frac{N(N-1)}{\Sigma n(n-1)}$	using $D = 1 - \Sigma\left(\frac{n}{N}\right)^2$	using $D = 1 - \frac{\Sigma n(n-1)}{N(N-1)}$	formula not specified		using $D = \frac{N(N-1)}{\Sigma n(n-1)}$
Understand the terms mean, median and mode — Calculate or compare the mean, median and mode of a data set, e.g. height/mass/size	1.2 7.4.1	✓	✓	✓	✓	✓	✓	✓	✓	✓
Find the interquartile range; understand percentiles	7.6.5							✓		
Use a scatter diagram to identify a correlation between two variables — Interpret a scattergram	7.14	✓	✓	✓	✓	✓	✓		✓	✓

Topic	Section	AQA	CCEA	Eduqas	Edexcel	OCR	WJEC	Cambridge pre-U	Edexcel International	IB
Make order of magnitude calculations	1.4.5	✓		✓	✓	✓	✓	✓	✓	✓
Select and use a statistical test — The chi squared test	7.20	✓	✓	✓	✓	✓	✓	✓	✓	
The student's t test	7.19	✓	✓	✓	✓	✓	✓	✓	✓	✓
The correlation coefficient	7.14 7.16 7.17	✓ test not specified		✓ test not specified	✓ Spearman rank correlation test	✓ Spearman rank correlation test	✓ test not specified	✓ Spearman rank and Pearson's linear correlation tests	✓ test not specified	
Understand measures of dispersion, including standard deviation and range	7.6	✓	✓	✓	✓	✓	✓	✓	✓	✓
Identify and determine uncertainties in measurements	2.1.5	✓		✓	✓	✓	✓	✓	✓	✓
Algebra										
Understand and use symbols: $=, <, \ll, \gg, >, \propto, \sim$	No exemplification required	✓		✓	✓	✓	✓	✓	✓	
Change the subject of, substitute numerical values into and solve algebraic equations	2.1 6	✓		✓	✓	✓	✓	✓	✓	
Use a logarithmic scale, e.g. growth rate of a microorganism, e.g. yeast	3.2 3.6.4	✓		✓	✓	✓	✓		✓	✓

Topic		Section	AQA	CCEA	Eduqas	Edexcel	OCR	WJEC	Cambridge pre-U	Edexcel International	IB
Graphs											
Translate information between graphical, numerical and algebraic forms	Understand that data may be presented in a number of formats and be able to use these data	3.6	✓		✓	✓	✓	✓	✓	✓	
Plot two variables from experimental or other data	Select an appropriate format for presenting data, bar charts, histograms, graphs and scattergrams	3.1-3.5	✓	✓	✓	✓	✓	✓	✓	✓	✓
Understand that $y = mx + c$ represents a linear relationship	Predict/sketch the shape of a graph with a linear relationship	3.6.2	✓		✓	✓	✓	✓			
Determine the intercept of a graph	Read an intercept point from a graph	3.6.2	✓		✓	✓	✓	✓	✓	✓	✓
Calculate rate of change from a graph showing a linear relationship	Calculate rate of change from a graph showing a linear relationship	3.6.3	✓	✓	✓	✓	✓		✓		
Use this method to measure the gradient of a point on a curve	Draw and use the slope of a tangent to a curve as a measure of rate of change	3.6.3	✓		✓	✓	✓	✓	✓		✓
	Recognise when to join points with straight lines and when to use a straight or curved best fit line; choose the line	3.5.5									
Geometry and trigonometry											
Calculate the circumferences, surface areas and volumes of regular shapes	Calculate the circumference and area of a circle, the surface area and volume of rectangular and cylindrical prisms and of spheres	2.2	✓		✓	✓	✓	✓	✓		

Index

addition 9–11, 14, 15, 16

allele frequency 110–112

antilog 20, 71–72

area 22, 34–40, 74–75, 85–86

area under the curve 118–119

average 114–115; *also see* mean; median; mode

axes 52–60, 62–64, 68–69, 71–74, 88

bar graph 54–57, 118, 119

BMI 13–14

BODMAS 16

calculator use 14–16, 20–21, 25–27

calorimetry 13

categorical data 52, 54–55, 129–130, 137, 141

cell cycle 75

centiMorgan 106

chi squared (χ^2) test 130, 141–144

co-dominance 95–96, 100

compensation point 64

concentration 19–21, 28–29, 40, 42–44, 53–54, 57, 60–61, 63–64, 68, 122, 130, 139–141

confidence limits 123–125

continuous data 55–56

correlation 102, 122, 128–136

critical values 125, 127, 133, 136, 138–141, 143–144

cube 36, 38, 80

cylinder 36, 39–40

data 11, 15, 25–27, 34, 44, 50, 52–62, 75, 87–88, 113–115, 118–121, 124–126, 128–131, 133–134, 137, 139–142, 144

decimal places 24–27

decimals 18, 24–27

degrees of freedom 123–125

dihybrid inheritance 96–104, 122, 124

discrete data (discontinuous data) 55

Disney's index 46

distribution – bimodal 117

distribution – normal 56, 115–116, 118, 127, 130

distribution – skewed 116–117

division 13–17, 19, 24, 34, 39, 47, 50, 67, 72, 81–82, 100, 113–114, 119, 139, 143–144

DNA 17, 40–42, 81, 91

ECG 73–74

ecological pyramid 87–89

energy flow diagram 12

epistasis 102–103

estimation 14–16, 34, 38, 44, 61, 64, 74, 87, 114, 119, 124–125, 130

fractions 13, 24–25, 40, 55, 100

frequency histogram 55–57

genetic code 17

genetic crosses 18, 22, 90, 95–96, 103–104, 141–142

genetic mapping 106–107

gradient 34, 63, 68–70, 73–74, 133–134

gross primary productivity (GPP) 11–12

haemocytometer 46–47

Hardy–Weinberg equilibrium 110–111

heterozygosity index 46

histogram 55–57, 118–119

hydrogen peroxide 44, 68

hypothesis 122–124, 127–129, 131, 133–134, 136–144

incomplete dominance 94–95, 100, 107

indices 14–19, 44–46

interval data 129–130, 137

kite diagram 74–75

lethal recessives 102

level of significance 123, 125, 133–134, 136–138, 140–141, 143

line graph 57–60, 118–119

linkage 104–107

Lincoln index 44–45

logarithm 18, 20, 29, 53, 58, 62, 72

magnification 79, 81–85

Mann–Whitney U test 130, 137–139

map unit 106

mean 15–16, 29, 56, 59, 67–68, 87, 114–120, 122, 124–126, 131, 137, 139–141

median 56, 114–117, 121, 137–138

microscope calibration 81–83

mode 56, 114–117

molarity 43–44

moles 19, 43–44

monohybrid inheritance 91–98, 101

multiplication 12–13, 15–16, 18–19, 24–25, 34, 50, 62, 67, 80–81, 85, 110–111, 113

negative numbers 9–11, 19, 27–29, 119

net primary productivity (NPP) 11–12

nomogram 75

non-Mendelian ratios 102–107

null hypothesis 122–123, 127–128, 131, 133–134, 136–144

Occam's razor 17

one-tailed test 127–128, 136

order of operations 16

ordinal data 129–130

oxygen dissociation curve 64–67

partial dominance 94–95

Pearson linear coefficient test 130, 134–137

percentage 25, 32–35, 42–43, 56–57, 64–67, 74–75, 110–111, 120

percentage error 34–35

percentile 120–121

per cent area cover 34, 74–75

per cent calculation 32

per cent concentration 42–43

per cent decrease 33

per cent frequency 34, 74, 110

per cent increase 33, 66

pie chart 75

PMCC 130, 134–137

population growth curve 21, 70–72

population pyramid 56–57

positive numbers 9, 10

powers 14, 16, 18–20, 53, 80–82, 85–87

probability 91, 93–94, 100, 104, 110, 113–114, 119, 123

proportion 12, 21–22, 34–42, 45, 55–56, 60–61, 68, 75, 87–88, 90, 92–93, 97–101, 105, 111, 114, 116–117

Punnett square 92–95, 97, 103–105, 107

pyramid of biomass 87–89

quadrat 34, 74, 85–87

quartile 120–121

range 14, 19, 52–53, 55–56, 58, 71, 79, 118–119, 121, 124, 129

ranking 126–127, 138

rate of reaction 52–54, 60–61, 63–64, 68

ratio 21–23, 36–40, 42, 90–107, 111, 122, 142, 144

recombination frequency 105–107

respiratory quotient (RQ) 13, 22–23

rounding 14–15, 26–27, 72

sampling 113–114

Sankey diagram 12

scale 21, 29, 34, 52–59

scale bar 81, 83

scale break 59, 62, 71, 73–75, 79–89

scatter graph 128–129, 133

serial dilution 19–20

sex linkage 107

SI units 80–81, 86–87

significant figures 27

Simpson's index 45–46

Spearman rank correlation test 130–134

species evenness 45

species richness (species diversity) 45

sphere 36, 39

spirometer 72–73

standard deviation 118–120, 122, 124–125

standard error 119

standard notation 14–16, 18

subtraction 9, 10, 11–12, 14, 15, 16, 19, 50, 66, 69, 71, 125

surface area 22, 35–40

t test 119, 124, 130, 139–141

tangent 69–70

test cross 93, 98, 100–101, 103

tidal volume 72–73

two-tailed test 127–128, 133

units 10, 19, 27, 29, 35, 42–44, 58, 62, 68, 80–87, 106

variance 119–120, 122, 124–125, 139, 141

vital capacity 72–73

volume 12, 22–23, 35–40, 42–44, 46–47, 65, 68, 72–73, 81, 86–87

water relations 10, 28